Advances in Combustion Technology

This edited volume on combustion technology covers recent developments and provides a broad perspective of the key challenges in this emerging field. Divided into two sections, the first one covers micro-combustion systems, hydrogen combustors, combustion systems for gas turbines and IC engines, coal combustors for power plants and gasifier systems. The second section focusses on combustion systems pertaining to aerospace including supersonic combustors, rocket engines and gel propellant combustion. Issues related to energy producing devices in power generation, process industries and aerospace vehicles and efficient and eco-friendly combustion technologies are also explained.

Features:

- Provides comprehensive coverage of recent advances in combustion technology.
- Explains definite concepts about the design and development in combustion systems.
- Captures developments relevant for the aerospace area including gel propellant, aluminium-based propellants, and gas turbines.
- Aims to introduce the combustion system in different industries.
- Expounds novel gasification systems with reference to pertinent renewable technologies.

This book is aimed at researchers and graduate students in chemical, mechanical and aerospace engineering, energy and environmental engineering, and thermal engineering. This book is also aimed at practicing engineers and decision makers in industry and research labs, and petroleum utilization.

Advances in Combustion Technology

Edited by Debi Prasad Mishra

CRC Press
Taylor & Francis Group
Boca Raton London New York

CRC Press is an imprint of the
Taylor & Francis Group, an **informa** business

First edition published 2023
by CRC Press
6000 Broken Sound Parkway NW, Suite 300, Boca Raton, FL 33487–2742

and by CRC Press
4 Park Square, Milton Park, Abingdon, Oxon, OX14 4RN

CRC Press is an imprint of Taylor & Francis Group, LLC

© 2023 Taylor & Francis Group, LLC

ISBN: 978-0-367-50152-5 (hbk)
ISBN: 978-0-367-50156-3 (pbk)
ISBN: 978-1-003-04900-5 (ebk)

DOI: 10.1201/9781003049005

Typeset in Times New Roman
by Apex CoVantage, LLC

Dedicated to my wife Itishree
and my son, Rutamvar

Contents

Contributors

V. Ganesan (late)
Department of Mechanical
    Engineering
Indian Institute of Technology
Madras, Chennai, India

Swarup Y. Jejurkar
Department of Space Engineering and
    Rocketry
Burla Institute of Technology
Mesra, Jharkhand, India

Sangeeta Kohli
Department of Mechanical Engineering
Indian Institute of Technology
Delhi, India

PK Ezhil Kumar
Dinesh Engineering Industries
Thiruvallur Dist, Tamilnadu, India

Sudarshan Kumar
Department of Aerospace Engineering
Indian Institute of Technology
Powai, Mumbai, India

Debi Prasad Mishra
Department of Aerospace Engineering
Indian Institute of Technology
Kanpur, India

Manisha B. Padwal
Department of Aerospace Engineering
Indian Institute of Technology
Jammu, India

P.A. Ramakrishna
Department of Aerospace Engineering
Indian Institute of Technology
Madras, India

V. Ramanujachari
Department of Aerospace Engineering
Indian Institute of Technology
Madras, India

Nikunj Rathi
Department of Aerospace Engineering
Indian Institute of Technology Madras,
Chennai, India

M.R. Ravi
Department of Mechanical Engineering
Indian Institute of Technology
Delhi, India

Mahesh S.
Department of Aerospace Engineering
Indian Institute of Space Science and
    Technology
Trivandrum, India

Saurabh Sharma
Department of Aerospace Engineering
Indian Institute of Technology
Powai Mumbai, India

Vijayashree
Department of Mechanical Engineering
Indian Institute of Technology
Madras, Chennai, India

Preface

Combustion has been part and parcel of human life since time immemorial. It is the most common and elegant method of converting the chemical energy stored in fuel to a useful form of energy—namely, thermal, mechanical, electrical and others—that can be used for several purposes particularly in modern life. In modern times, life without use of external energy is unthinkable as modern people use a large number of gadgets for leading a comfortable life. However, blatant uses of combustion systems for harnessing useful and deployable energy for making all gadgets operable in an uninterrupted manner has created several problems: an energy crisis and environmental pollution. In order to sustain life at the same pace, combustion technology is likely to continue to play a major role in driving the materialistic developmental pace as per the demand of the modern system. Hence it is important and essential for us to design and develop combustion driven devices that can be more efficient and effective in abating environmental pollution, which has threatened the very survival of life on this beautiful blue planet. The emissions from combustion systems can be minimized by adopting two methods— namely, redesigning new combustion systems and retrofitting existing combustion systems. Efforts have been made by researchers over several years to design and develop several innovative combustion concepts—namely, inverse jet flame, flameless combustion, trapped vortex combustion, lean premixed combustion, partially premixed flame, fuel-rich-fuel-lean combustion. Besides this, there is another way of substituting fossil fuel with fuel-like hydrogen and biomass that can not only make combustion systems efficient but also environment friendly. Besides this, gel propellants and aluminized fuel-rich propellants are some of the advanced combustion systems that have been designed and developed by researchers. Based on this background this compendium covering advanced technological aspects of combustion systems is designed as an edited book which is expected to be helpful for post-graduates and researchers in the field of mechanical, aerospace, chemical and other engineering disciplines. It consists of ten chapters written by well known researches in the field of combustion technology.

Chapter 1 deals with microcombustors, which are considered to be next-generation combustion applications—namely, power MEMS, micropropulsion systems, microreactors and other devices. Of course combustion at micro-length scale is primarily driven by the behaviour of flames in continuous thermal contact with the channel walls and a competition between the heat released by combustion and heat losses to the ambient. In Chapter 1, an annular cylindrical microcombustor is designed and developed along with its characterization. This chapter covers important results on the proposed flame stabilization mechanisms, flame structure, and stability limits obtained with the help of numerical analysis of the microcombustor. Subsequently Chapter 2 deals with inverse jet flame based swirl combustor which is emerging as a substitute for diffusion flame based combustors in industry to meet the new norms of environmental regulations. In Chapter 2, the feasibility of employing a coaxial inverse jet flame (IJF) as an alternative technology for adaption in gas turbine combustors is being considered to overcome the safety issues associated with the lean premixed combustion. This system can alleviate the problems of flame flashback and lean blowout issues of lean premised flame combustion systems as IJF has the ability to control the air and fuel jets independently. A swirl IJF based dump combustor has been investigated experimentally in terms of visible appearance, flame height and flame stability along with emissions as reported in detail in this chapter.

Chapter 3 deals with state of the art in burner technology for hydrogen fuel. A novel burner design is considered for this technology which indicates promising results. In order to overcome issues with hydrogen storage, several technologies for combustion are discussed for the future, which is essential for successful deployment of hydrogen based combustion systems. Chapter 4 is concerned with design and development of a novel flameless combustor of liquid fuels to achieve ultra-low NO_x emissions from combustion systems employed for various applications such as industrial furnaces and gas turbines. This combustor, with the help of a new air injection method for kerosene, can achieve thermal intensities varying from 10–20 MW/m^3 with a maximum reactant dilution ratio, R_{dil}, of 3.75 and almost zero NO_x emissions.

Subsequently in Chapter 5, both non-conventional Spark Ignition (SI) and Compression Ignition (CI) engines that can improve combustion and reduce emissions are discussed extensively. The details about four non-conventional engines, Gasoline Direct Injection (GDI), Common Rail Direct Injection (CRDI), Homogeneous Charge Compression Ignition (HCCI) and Stratified Charge (SC) engines, are covered in this chapter. How the design method of biomass gasifier systems can convert solid biomass into a combustible gas, called *producer gas* or *synthesis gas* (*syngas*), is covered in Chapter 6. The design process for downdraft biomass gasification systems is based on the extensive experimental data of Swedish researchers over several decades nearly a century ago, and the empirical approaches are presented here. The design process for downsizing to micro-sized gasification systems as low as 1 kg/hr or less and that can be used for domestic cookstove applications is discussed. Chapter 7 deals with in-depth studies of a 2D trapped vortex combustor along with various aspects of this combustor. The optimum size and shape of its cavity based on flow physics, pressure drop and fuel-air mixing are brought out to achieve good fuel-air mixing and a stable flame (wider lean blow out (LBO)) limits. This combustor is characterized in terms of factors affecting the combustion efficiency, the mainstream velocity, level of mainstream premixing, cavity equivalence ratio and injection momentum.

The next three chapters are devoted to important combustion technology in the field of aerospace engineering. Chapter 8 deals with supersonic combustion ramjet (scramjet) technology which is being pursued vigorously by some of the advanced countries. Though the fundamental aspect of this combustion system looks to be simple and straight forward, its realization is a technological challenge. The scramjet engine consists of an air intake, isolator, combustion chamber and nozzle. This chapter highlights the supersonic combustion and propulsion system, and the technological challenges of making advancements in the field of scramjet engines.

Chapter 9 presents a brief account of advances in combustion technology for gel propellants, which are considered to be the next-generation propulsion system. The essential elements of gel propellant combustion involving the propellant development, characterization, atomization, and combustion are covered in this chapter. Chapter 10 deals with experimental results of an aluminized fuel-rich propellant having energetics similar to boron based fuel-rich propellant which is considered to have zero residue, high density and higher burn rates. This study indicates that fuel-rich propellants made with higher loading of aluminium (30% by weight) has resulted in zero residue.

Several fellow researchers in the field of combustion—namely, Prof. P.A. Ramakrishna, Dr. Nikunj Rathi of IIT Madras, Dr. Manisha B. Padwal of IIT Jammu, Dr. V. Ramanujachari of IIT Madras, Dr. P.K. Ezhil Kumar, Prof. M.R. Ravi and Prof. Sangeeta Kohli of IIT Delhi, Prof. V. Ganesan and Dr. Vijayashree of IIT Madras, Prof. Sudarshan Kumar and Mr. Mahesh S., of IIT Bombay, Dr. Swarup Y. Jejurkar of BITs Mesra and Dr. S. Mahesh of IIST Trivandrum—have contributed chapters for this edited book and I am grateful to them. I am also indebted to all these authors and my students and other staff members, Shiv Kumar, Dr. Pravendra Kumar, Dr. H. Vishnu, Abhishek, Imran, Assiz and Utpal who have indirectly helped me with this work. The patience and perseverance, support and follow-up by Dr. Gagandeep Singh of Taylor and Francis, India is highly appreciated. Last but not the least, I am grateful to the support and patience displayed by my family members during this project without which it could not have fructified.

Editor's Biography

Dr. Debi Prasad Mishra, a professor in the Department of Aerospace Engineering and Design Programme, Indian Institute of Technology Kanpur (IITK) Kanpur, India, is currently working as the Director of the National Institute of Technical Teachers Training and Research (NITTTR), Kolkata. He had held the Indian Oil Golden Jubilee Professional Chair in IIT Kanpur. He has been conferred with several awards and recognitions. His major areas of research interests are propulsion, combustion and atomization. He has written five textbooks and six edited books in the field of combustion and propulsion. He has published 238 research papers in international/national journals and has five patents to his credit. He has undertaken more than 37 research and consultancy projects from various agencies, research labs and industries. He has developed six online courses under the Swayam platform, a national initiative of MHRD. Besides this, he has designed and developed a unique MOOC course, "Introduction to Ancient Indian Technology" on the Swayam platform of NPTEL. He has written several articles to popularize science and technology among common people and delivered several lectures and discussions on TV and radio related to ancient science and technology and other different topics. He is well known as a motivational speaker and is very popular among youth circles.

1 Annular Microcombustor and Its Characterization

Swarup Y. Jejurkar and Debi Prasad Mishra

CONTENTS

1.1 INTRODUCTION

Combustion in narrow confined spaces is constrained by flow and heat transfer within the equipment and the solid structure surrounding the flame. A familiar example of this is found in the miners' lamp consisting of wire gauze with openings smaller than the "quenching diameter", which prevents explosion of methane-rich gases. Although, not strictly defined, quenching diameter varies in the range 600–3000 μm [1]. Incidentally, combustion in small-bore tubes is also used for estimating the quenching diameters of fuels. Microcombustion occurs at length scales

DOI: 10.1201/9781003049005-1

1

near the quenching diameter of fuels and usually the length scales of channels used for microcombustion are $\sim O(1)$ mm or smaller [2]. Combustion occurring in the cracks of internal combustion engines is also a common example of microcombustion. Power sources based on microcombustion are potentially useful in cases where combustor length scale cannot be far removed from the quenching diameter.

This chapter is devoted to a discussion of the fundamental challenge in microcombustion, a strategy for heat augmentation devised to address the challenge, and a description of the microcombustor designed to realize this strategy using hydrogen. Discussion is focused on the design of microcombustor, a numerical model to analyze the configuration, and parametric studies using the model.

1.2   OVERVIEW OF MICROCOMBUSTION

This section presents an overview of microcombustion using its status as a source of portable power among other competing methods and scope of applications. Methods of fabrication are introduced to help understand some of the technological aspects discussed in later sections.

1.2.1   PORTABLE POWER SOURCES

There is a growing trend in aviation, biotechnology, chemical process industry, information technology, and medicine towards the miniaturization of devices and processes—viz., micro air vehicles, substrate characterization, reactor-separators, communication devices, and drug-delivery systems. This trend is fuelled by the need to squeeze maximum out of what is available and to cut costs, and also since the miniature devices offer better control, more precise measurements, and more flexibility than the conventional ones. New applications also demand greater sophistication in terms of efficiency, weight, and reliability; for example in volume-constrained space propulsion and battery-operated devices.

Methods such as electrochemical cells, fuel cells, and most importantly, combustion, which converts chemical energy into thermal energy, provide the necessary energy for operating portable devices. Combustion is an important route to obtain thermal energy from various fuels including hydrogen, natural gas, fuel oils, gasoline, and alcohols. Thermal energy then transforms into kinetic or electrical energy via subsidiary routes. Due to the restriction on available volume, combustion must take place at small length scale in miniature devices. In this respect, it is termed as microcombustion because the characteristic length scale of the confined space is about 100–1000 μm. A successfully miniaturized or "microcombustor" would be integrated with other miniature devices including compressors, turbines, pumps, and valves to achieve operational flexibility, higher operating efficiency, reduction in environmental hazard, and improvement in economy and reliability. A microcombustor differs from a conventional gas turbine combustor, Li-ion battery, and fuel cell in a number of aspects as summarized in Table 1.1.

Striking contrasts in length scale (l_c), surface area per unit volume (SAV), thrust-weight ratio, and combustion efficiency (η_{comb}) are apparent from Table 1.1.

TABLE 1.1
Comparison of Competitive Miniature Energy Converters and Gas Turbine Combustors.

Variables	GT Combustor [3]	Microcombustor [3]	Li–Poly Cell [5]	SOFC [6]
l_c	~20 cm	0.05–1.0 cm	NA	1 cm
SAV	small (~4)	large (~200)	NA	4
P_C	~40 atm	0.01–4.0 atm	normal	normal
T	~1600 K	~1600 K	normal	~1073 K
$\dot{m}$	~140 kg/s	0.01–2 g/s	–	H_2: 0.5, air: 20 µg/s
τ_{res}	~7 ms	~0.5 ms	–	–
thrust/wt	20:1 [4]	100:1 [4]	–	–
η_{comb}	>99.9%	~70–80%	–	~45–55%
P	143 MW	~10 W	187 Wh/kg (3.7 V)	27 W (0.7 V)

The small size of microcombustors, while necessary from application point of view, also gives rise to very large *SAV* and inadequate residence time (τ_{res}). Consequently, their combustion efficiency is very poor in comparison to the conventional gas turbine (GT) combustor. On the other hand, obtainable thrust-weight ratio (based on a complete micro gas turbine) is higher for a microcombustor since this ratio scales inversely with l_c [3]. Although power output of microcombustor (*P*) is low, it could increase simply by scaling out (i.e. by employing arrays of microcombustors).

Self-sustaining microcombustion is difficult due to heat losses through walls that could eventually thermally quench the flame. Heat losses are mainly due to a large *SAV*. Additionally, residence time necessary to complete mixing and combustion of reactants may not be easily available in a microcombustor, causing incomplete combustion and poor operating efficiency. Apart from flame quenching induced by heat losses, radical loss to reactive walls (chemical or radical quenching) can also occur in microcombustion.

An alternative strategy to use fuels at small length scales is catalytic combustion, a heterogeneous process. Both catalytic and homogeneous microcombustion routes are being explored [7]. In catalytic combustion, noble metals like Pt, Rh, and Pd supported on alumina (Al_2O_3) and zirconia (ZrO_2) are coated on to the microcombustor wall to lower the activation energy of reactants so that reactions occur at lower temperature. Reaction rate in catalytic combustion also scales with the specific surface area (available catalyst surface area per unit geometric surface area) that increases enormously due to the porous supports.

1.2.2 APPLICATIONS OF MICROCOMBUSTION

One important application for microcombustion is in military communication devices. Many more areas have been identified, as illustrated by Figure 1.1.

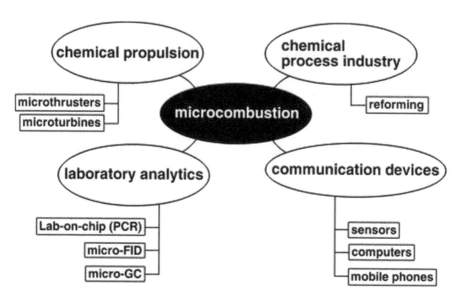

FIGURE 1.1 Potential applications of microcombustion.

As also evident in Table 1.1, microcombustors could replace batteries in many devices because the best available Li-ion batteries have energy storage capacity of 1.5 MJ/kg against 45 MJ/kg of liquid hydrocarbons [8]. Effectively, poorly performing microcombustors might be capable to offer competitive advantage over the best batteries.

Consumer electronic products like mobile phones, laptop computers, as well as military communication devices can benefit from this characteristic. Efficiency of microcombustors is an important factor when the possibility of replacing batteries by microcombustors is considered. Minimum level of efficiency should be decided in terms of the overall system efficiency; for example, the power cycle in which microcombustor is used determines its efficiency. Additionally, one of the main hurdles is the technology to convert thermal energy from microcombustion into useful electrical power. Thermoelectric (TE) and thermophotovoltaic (TPV) elements are useful for this conversion although their conversion efficiencies (~10%) are not high enough.

Another application of microcombustion is in the propulsion of small air vehicles with mass less than 10 kg and thrust requirements in 1–10 mN range. Thrust augmentation may be possible by employing arrays of microcombustors and thrusters. Better control and improved redundancy are the two major advantages of such an arrangement. This scale-out or series-parallel arrangement enhances power output and thrust obtainable from devices that individually give small outputs. Thrust vectoring may also be possible with microcombustors. Thus, they can be potentially useful in propulsion systems, although electric propulsion systems currently compete with microcombustion in this area. Microturbomachinery built by replicating a conventional gas turbine cycle and employing microcombustion has also been tested

[4]. This approach relies on fabrication techniques used in micro-electro-mechanical systems (MEMS). The focus in this chapter is on applications in portable power generation and propulsion.

In order to harness thermal energy from microcombustion, it is necessary to provide means for conversion into usable electric or kinetic energies. Conversion to kinetic energy is achieved by conventional turbines [9] and thrusters [10] miniaturized specially for this purpose, while conversion to electric power is obtained by TE [11] or TPV [12] effects. Associated conversion elements like heat spreaders and circuitry are also miniaturized to keep the overall package size within prescribed limits. In a TE generator, heat of microcombustion passes through heat spreaders out of the combustor and resulting thermal gradients are used to set up voltage as per the Seebeck-Peltier effect. Part of the heat is rejected to the atmosphere via radiator fins and other compact heat exchanger elements. The module may also contain an inverter. In a TPV generator [12], a selective emitter or a gray body emitter-band pass filter combination captures energy photons obtained from combustion. Photovoltaic cells absorb the emitted radiation and convert into electric power through circuits. These two effects are especially attractive to microcombustion-based generators because the converters have no moving parts and can fit in a small volume.

1.3 THE FUNDAMENTAL CHALLENGES

Miniaturized devices need small amounts of power commensurate with their size and energy utilization rate. Considering their size, the available volume for power plants is small and this reduces the space available for combustors to < 1 cm³ [3]. Linear dimension of the combustor is further constrained and is about a few mm or less. The combustor length scale is incidentally of the same order of magnitude as the reaction zone thickness (and quenching diameter) for most of the practical fuels. This leads to narrow flammability limits, thermal quenching of the flame, and difficulty in flame stabilization within a microcombustor [13].

1.3.1 RELEVANT LENGTH AND TIME SCALES

Simple order-of-magnitude analysis can be used to further explain the far-reaching effects of small length scales. As shown in eq. (1), the ratio of heat loss rate to heat release rate is inversely proportional to the hydraulic diameter of the combustor so that reduction in combustor size would lead to thermal quenching and narrowed flammability limits.

$$A_s \sim D_h^2 \text{ and } V \sim D_h^3, \text{ which lead to} \frac{A_s}{V} \sim \frac{1}{D_h}$$

$$\Rightarrow \left(\dot{q}_{loss}'' A_s / \dot{q}'' V \right) \sim \left(1/D_h \right) \tag{1}$$

Similarly, eq. (2) shows that flame stabilization within the microcombustor might be prevented by inadequate residence time, which turns out to be directly proportional to the hydraulic diameter during constant velocity scaling.

$$\tau_{res} = \frac{\rho V}{\dot{m}} = \frac{\rho V}{\rho A_c u_{in}} \sim \frac{D_h^3}{D_h^2} \sim D_h \quad (\because u_{in} = \text{constant}) \qquad (2)$$

All microcombustors must be designed and operated with proper consideration of these restrictions on the time and length scales. The restrictions have been circumvented by some encouraging developments in this field.

1.3.2 CURRENT CHALLENGES

Preheating of fuel-air mixture by thermal feedback originating from hot product gases is a plausible flame stabilization mechanism at millimeter and micrometer scale. Heat transfer processes taking part in thermal stabilization however are not studied in detail. The effectiveness of the flame holding mechanism, and ratio of heat loss–heat recirculation are critical for enhanced flame stability. Studies analyzing the detailed flame structure acquire significance for the fundamental understanding of microcombustion and effects of high velocity and heat losses on flame stability.

In experiments, it has not been possible to rely solely on internal heat recirculation for achieving flame stabilization. Resulting combustor configurations are difficult to build and integrate with other components as complicated structural arrangements are devised for controlling heat losses (by using elaborate heat exchanger arrangements) and flame stabilization. Due to the presence of extra solid structure, overall weight of the assembly is likely to be larger and pressure losses resulting from narrow tubes and passages in porous materials may be higher than tolerable. More importantly, too much solid structure delays the onset of steady state and lengthy start-up is inevitable.

Cylindrical microcombustors have been studied extensively, both experimentally and numerically. Their shape factors are also suitable for propulsion and power generation applications. Close thermal contact with other components would provide a path to heat losses by conduction and/or radiation. Consequently, the possibility that flame is always subject to heat losses to the walls in microcombustion should be factored into microcombustor design. However, instability and thermal quenching of flame can be addressed by a careful structural and thermo-fluidic design based on earlier experience with the heat recirculating microcombustors.

There is a need to devise and study a heat recirculating microcombustor configuration that addresses these challenges and shortcomings at the micro- and mesoscale. It is possible to obtain a better configuration if means are provided for improving the thermal isolation of a part of a microcombustor and stabilize the flame near this region.

1.4   THE ANNULAR MICROCOMBUSTOR

1.4.1   CHOICE OF CONFIGURATION

Annular configuration is advantageous in microcombustion-based power plants because it can be easily packaged in the limited space available for power plants in small-scale devices. In this configuration, heat releasing combustion reactions occur in an annulus surrounding the hollow core. Theoretical considerations involved in the proposed microcombustor concept are described next.

1.4.2 THERMAL ISOLATION

A major disadvantage of using solid rod in the flame zone is its tendency to store thermal energy, as its volumetric heat capacity, ρC_p, and thermal inertia, $(k\rho C_p)^{1/2}$, are high. Due to this property, solid temperature would increase slowly until thermal equilibrium is attained. The amount of heat absorbed before reaching a certain temperature will also be large. As such, external heating of the combustor is necessary prior to actual combustion and start-up from the cold conditions is very difficult. This is especially important in the start-up phase of combustion when more heat is required near the flame zone, or in other words, temperature near the flame zone must be high enough to allow flame development from the ignition event onwards.

In the annular microcombustor, partial thermal isolation of the core is achieved indirectly by artificially maintaining its walls at high temperature. For this, a hollow inner tube filled with an inert gas is used as the core of the combustor, as shown in Figure 1.2. Comparison of thermal properties of gases and typical solid materials in

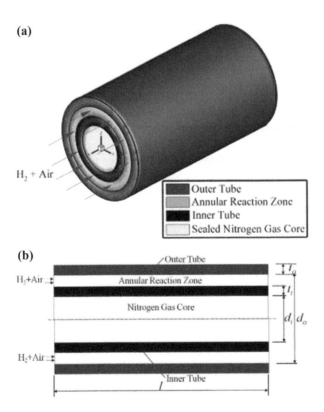

FIGURE 1.2 Schematic representation of annular premixed flame microcombustor; (a) 3-D view, (b) 2-D view.

TABLE 1.2
Comparison of Thermal Properties of Some Inert Gases and Solid Materials [14–16].

T (K) 300/1000	k (mW/m K)	ρ (kg/m³)	C_P (kJ/kg K)	α ($\times 10^4$) (m²/s)	ρC_P (kJ/m³ K)	$(k\rho C_P)^{1/2}$ (kJ/m² s$^{1/2}$ K)
He	156/360.6	0.1604/0.0481	5.193/5.193	1.873/14.466	0.833/0.249	0.361/0.299
Ar	17.84/43.58	1.603/0.4803	0.5215/0.5204	0.213/1.743	0.836/0.249	0.122/0.104
N_2	25.97/65.36	1.123/0.3368	1.041/1.167	0.222/1.663	1.169/0.393	0.174/0.160
CO_2	16.79/70.57	1.773/0.5292	0.8525/1.234	0.111/1.081	1.511/0.653	0.159/0.215
ceramic	1.75e03	2872	0.910	0.007	2613	2.1386
steel	16.25e03	8030	0.5025	0.0403	4034	8.0974

Table 1.2 shows that volumetric heat capacity and thermal inertia of gases are much smaller than solids and gases would reach thermal equilibrium much faster (temperature rises with a small amount of heat absorption). It is expected that such an arrangement would store the heat released in combustion during the flame development stage that would otherwise be lost and assist in the formation of a stable flame structure. As a result, lengthy external heating is reduced and might be entirely avoided.

1.4.3 Thermo-Fluidic Design

The gas in the core should be chemically inert at high temperatures. It should have low thermal diffusivity, a reasonable thermal storage capacity per unit volume, and it should be easily available and safe to handle.

Low thermal diffusivity of the gas in the inner tube would confine the heat absorbed from the incipient flame to regions near the walls of the tube containing the gas and increase the possibility of self-sustained combustion. In this respect, the decreasing order of thermal diffusivities for the inert gases is He>Ar>N_2>CO_2 at higher temperatures [14]. Table 1.2 summarizes the data of thermal diffusivity, volumetric heat capacity, and thermal inertia for the gases at $T = 300$ K and 1000 K and $P = 0.1$ MPa.

N_2 will prevent heat losses from regions near the walls during flame development better than either He or Ar due to smaller thermal diffusivity at high temperature. Additionally, a minor advantage in heat storage per unit volume would result by virtue of higher volumetric heat capacity (ρC_P). Although CO_2 is superior to N_2 in all the listed criteria, it is not easily available and hence N_2 was selected for storage in the inner tube core. N_2 will be charged and stored in the inner tube under pressure and undergo a constant volume heating process of a control mass during combustion since the tube walls are rigid. Calculations showed that the rise in N_2 pressure during the heating phase would be ~2% of the initial pressure in the inner tube. Thermal conductivity of N_2 is a very weak function of pressure and specific heat variation would be likewise negligible for the pressure change of this order. Additionally gas density, $\rho = m_{gas}/V_{it}$, is constant for a given initial pressure since walls are rigid and no mass is exchanged. Thus, thermal diffusivity will remain constant during the heating process and the pressure rise is not expected to couple with heat transfer across the inner tube.

1.4.4 GEOMETRIC DETAILS

Combustion reactions in the proposed microcombustor would occur in an annular cavity formed by the two concentric tubes, as shown in Figure 1.3a. The inner tube would contain N_2 sealed within and the outer tube would form the external surface of the microcombustor. Typical dimensions of these structural features are summarized in Table 1.3 and could be identified from Figure 1.3b. With these specifications, the annular zone is 1 mm wide in radial direction. The overall combustor volume is $1.9e-6$ m³ and surface area per unit volume based on the outermost wall surface area is 364 m⁻¹.

Reactants and products flow co-current to each other. Simple construction and unobstructed flow of reactants should result in smaller pressure losses under comparable conditions. Additionally, the resulting surface area per unit volume is much smaller in comparison to other configurations using a porous inert medium for flame stabilization. A simple conceptual design of a heat recirculating annular microcombustor is thus proposed on the bases of previous experiences with similar designs and thermal arguments supporting the use of a gas filled hollow inner tube, instead of porous solid rod.

1.4.5 VOLUME OF INNER TUBE

Table 1.2 clearly shows that thermal inertia of solid being higher than the gas, development of a "strong" wall boundary condition as early as possible is aided only by stemming the heat losses through the surrounding walls. From a thermal point of view, it is thus sufficient to create a low thermal conductivity region for preventing further loss of heat absorbed by the walls of the inner tube from the flame. Walls should be as thin as possible, subject to structural requirements and machining method. A too thick inner tube wall (very small gas volume) is not desirable as well

TABLE 1.3
Typical Component Dimensions of Annular Microcombustor.

Component	d (m)	t (m)	l (m)
inner tube (i)	0.007	0.001	0.02
outer tube (o)	0.011	0.001	0.02
annular zone (a)	0.002ᵃ	0.001	0.02
gas core (c)	0.005	—	0.02

Note: ᵃ Hydraulic diameter.

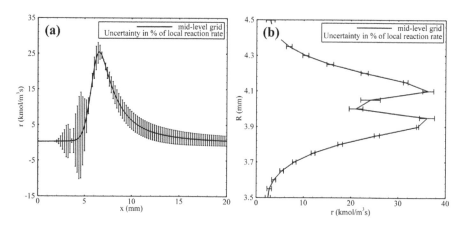

FIGURE 1.3   Reaction rate profiles for mid-level grid: (a) longitudinal, (b) radial with error bars for discretization uncertainty (case: V_{in} = 15 m/s, T_{in} = 300 K, ϕ = 1.0).

since it would absorb more heat and is likely to delay the heating of gas contained in the hollow core region. Using a low thermal conductivity solid instead of gas is not practical as the temperature might exceed service-ceiling temperature and lead to phase change.

1.5   MATHEMATICAL FORMULATION AND SOLUTION METHOD

The characteristic length scale of the annular microcombustor considered here is much larger than the limit for a continuum regime prescribed by Knudsen number (Kn), which is ~0.001 for l_c = 100 μm. As such, fluid flow is within the continuum regime as the mean free path of its molecules is much smaller than the combustor length scale. Hence, Navier-Stokes equations govern all flow processes in the micro-combustor. The annular microcombustor could be analyzed using a mathematical formulation of the reactive flow problem consisting of the governing equations, constitutive equations for material properties, thermal radiation, and finite rate chemical kinetics [17].

   The mathematical model of the annular microcombustor describes reacting laminar flow subject to variations of fluid and solid phase physical properties with temperature and composition [17]. Sub-models for thermal radiation and multicomponent mass transfer are included to account for their effects on heat losses and mass transport. In accordance with the known effects of Soret diffusion in hydrogen flames, species transport includes multicomponent Soret diffusion (for *all* the light and heavy species of the mixture) in addition to multicomponent Fickian diffusion in both mass and energy transport equations [18]. Radiation within the walls, between the walls, and within the gas phase is included [17]. Energy transport in the solid phase is also modelled explicitly since the length scales of the wall and reaction zone are similar in microcombustion [17]. The model equations are solved by a coupled density-based solver using at least second-order accurate discretization on a square non-uniform grid system [15]. Detailed estimates of numerical accuracy and uncertainty as well as model validation are obtained to establish the reliability of results. An axisymmetric computational domain was chosen for the characterization of annular microcombustor, as shown in Figure 1.2b. Finite-volume square cells discretized the computational domain. All computations were started with the deposition of heat in a selected circular zone located near the inlet plane of the annular section.

Sample results are shown for the stoichiometric hydrogen-air mixture (equivalence ratio, $\phi = 1$) computed using single step kinetics model (SSKM) of Hsu and Jemcov [19] for inlet velocity, $V_{in} = 15$ m/s (constant velocity profile), wall thermal conductivity, $k_{wall} = 1.75$ W/mK, and inlet temperature $T_{in} = 300$ K. The results indicate that a mid-level grid system consisting of 44600 cells might be adequate for the present purposes since its results follow closely with those of the fine grid having 146039 cells. Numerical uncertainty is expressed in terms of grid convergence index (GCI). Maximum computed uncertainty at the mid-level grid is 12.4% (± 0.21 kmol/m^{3}s) for reaction rate (r) in axial direction, as shown in Figure 1.3a. Although average uncertainty at the mid-level grid is higher than the fine level grid, the mid-level grid is computationally economical and limited the averaged extrapolated errors to values comparable with the fine grid solution. The maximum uncertainty at mid-level grid was 2.1% (± 0.49 kmol/m^{3}s) for reaction rate across the annular reaction channel. Comparison with the fine grid showed that the mid-level grid is suitable for computations.

Results of model validation tests are summarized in Figure 1.4 for the previous test cases. The ability of SSKM to predict the average wall temperatures is observed

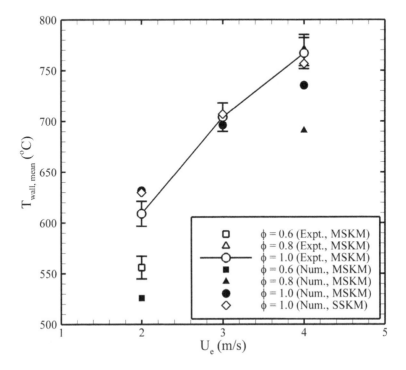

FIGURE 1.4 Summary of validation studies based on experimental data of Li et al. [20].

in the figure for the stoichiometric case as the predicted values are either within the experimental uncertainty or not far from the error bounds. The predicted values also compare favourably with a multi-step kinetics model (MSKM) [18].

The verified numerical solution method and validation of the computational model could be reliably used for investigation of the flame stabilization mechanism and parametric effects.

1.6   FLAME STABILIZATION MECHANISM

Flame stabilization within the microcombustor is an important design goal because a robust flame stabilization mechanism improves flame stability and enlarges the operational envelope. In microcombustion, a flame is stabilized by maintaining high wall temperatures or by using flow reversal, as in dump microcombustors and can combustors. While the traditional methods are well tested at the conventional scale, in microcombustion, they have not been able to provide flame stabilization for a wider range of operating conditions. Purely thermal feedback from walls (as against flow recirculation) could also provide wider stability limits. However, walls of a microcombustor play a dual role, acting as a conduit for heat losses and as a facilitator for flame stabilization. Besides this, a number of parameters, including inlet velocity magnitude and profile, pressure gradients near the flame front, velocity gradients near the wall, and wall thermal state, affect the flame stabilization in microcombustion. Mechanism of flame stabilization in the annular microcombustor is reported using the geometry specifications of Table 1.3.

1.6.1   THERMAL ISOLATION OF INNER TUBE

In order to understand flame stabilization process and the role of the inner tube in flame stabilization, time accurate simulations using single step kinetics model of Hsu and Jemcov [19] were performed for a representative case (V_{in} = 15 m/s, T_{in} = 300 K, k_{wall} = 1.75 W/mK, and ϕ = 1.0). Since the focus was on temperature-time relationship in the walls, a constant time step (0.001 s) much larger than the characteristic reaction time was specified.

Instantaneous contours of temperature in Figure 1.5 show that the heat released in an annular reaction zone moved predominantly in the downstream direction with time. During the transient process of flame movement towards upstream region, the inner tube wall and N_2 gas also heated up, while the outer wall was partially cold in the upstream zone (Figure 1.5b). The inner tube gradually heated up as it stored a part of the heat released by combustion and its temperature increased to become nearly equal to the post-flame zone temperature at the steady state (Figure 1.5c). Note that the wall thermal conductivity (k_{wall} = 1.75 W/mK) was sufficient to conduct the heat near to the reactor inlet and the flame consequently stabilized near the inlet. The thermal feedback mechanism established during the transient conditions depicted in Figure 1.5a and 1.5b consisted of convection from hot post-flame gases to the combustor walls, axial heat conduction from the outlet end towards the inlet end of walls, and convection from walls at the inlet end to the incoming reactant mixture.

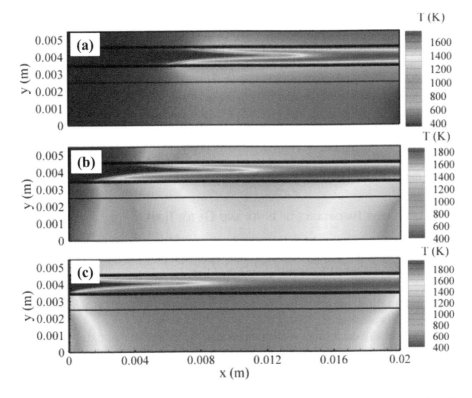

FIGURE 1.5   Instantaneous contours of temperature (K) across the domain; (a) $t = 2$ s, (b) $t = 8$ s, (c) steady state. (Operating conditions: $V_{in} = 15$ m/s, $T_{in} = 300$ K, $\phi = 1.0$, $k_{wall} = 1.75$ W/mK.)

On the other hand, heat transfer near the flame zone was multi-dimensional and complicated by the heat transfer from two branches to the nearest walls. Temperature of the inner tube wall increased at a faster rate than the outer tube wall because N_2 gas confined the absorbed heat in the near wall region during the warm-up period due to its low thermal conductivity (Figure 1.5a). Consequently, the inner tube temperature was higher than the outer tube at steady state. Thus, the inner wall heat conduction transferred more heat in the upstream direction and preheated the cold reactant mixture more than that of outer wall.

The conductive heat losses cannot be eliminated in microcombustion since the microcombustor is in a close thermal contact with the compressor, turbine, and other components of a micro gas turbine engine. Additionally, heat losses in microcombustion influence the stability of flame because they could cause local or global extinction. However, the inner tube was effectively isolated from the surrounding environment since the flame enveloped the tube from its outside and N_2 thermal conductivity was low (0.0782 W/mK at 1200 K) in comparison to the wall material. In other words, the inner tube acted as a miniature thermal reservoir for supplying heat to the reactants and stabilized the flame.

According to the Lewis-von Elbe criterion [1], the flame flashback would occur if the fluid velocity gradient at the wall were less than the ratio of local flame speed and the quenching distance. However, this criterion applies to cold walls only; while the wall temperatures were above the self-ignition temperature of H_2-air stoichiometric mixture (~783 K) the flame reached the wall without quenching in the present case. A preliminary analysis of the flow field in the pre-flame zone was carried out to further investigate the velocity and pressure field as suggested by Lee and T'ien [21]. The streamlines converged in the pre-flame region, which indicated that the flow did not retard. Hence, for the present case, the thermal state of the wall determined the flame stabilization and the flame stabilized at a point where heat conduction balanced the heat losses occurring near the inlet and walls.

1.6.2 RELATIVE IMPORTANCE OF INNER AND OUTER TUBES

The lower half of Figure 1.6 schematically shows the heat transfer paths and the mechanism of heat transfer to the reactants. Although the outer tube wall lost heat to the ambient (Figure 1.5), it still contributed heat for preheating the cold reactant mixture. The wall conduction occurred in longitudinal and transverse direction, while wall-to-fluid convective heat transfer in the transverse direction achieved preheating in this combustor.

Both inner and outer tubes recirculate heat to the cold reactant mixture. In order to elucidate their relative importance at different operating conditions, total surface heat fluxes at the wall surfaces between inner tube and gas mixture as well as between outer tube and gas mixture are plotted in Figure 1.7 for two extreme cases of k_{wall} (Figure 1.7a) and V_{in} (Figure 1.7b). Both convective and radiative components are included in the heat flux magnitude. A positive sign indicates heat transfer from the fluid to the wall region and a negative sign indicates heat transfer in the opposite direction (heat reflux). Accordingly, both inner and outer tubes transfer heat to incoming cold reactant mixture in the pre-flame region, while hot product gases transfer heat to surrounding walls in the post-flame region.

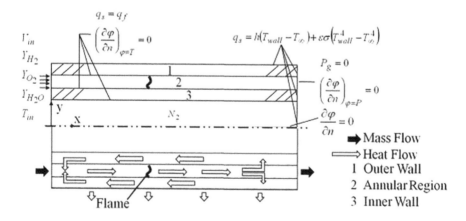

FIGURE 1.6 Schematic of thermal feedback mechanism in annular microcombustor.

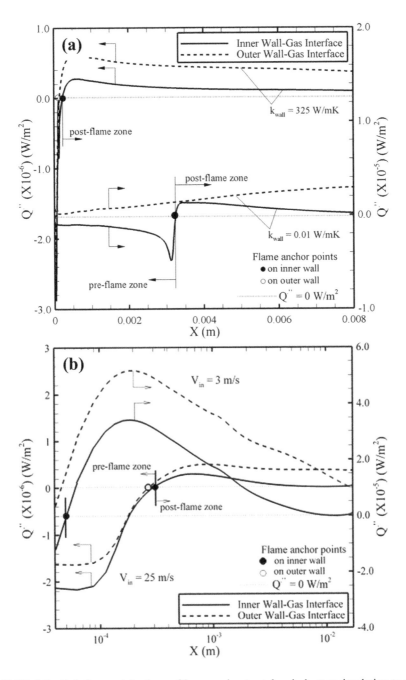

FIGURE 1.7 Relative contributions of inner and outer tubes in heat recirculation to cold fuel-air mixture in terms of total surface heat flux; (a) k_{wall} = 0.01 and 325 W/mK (V_{in} = 15 m/s), (b) V_{in} = 3 and 25 m/s (k_{wall} = 1.75 W/mK). A Negative sign indicates heat transfer from wall to fluid (heat reflux).

Figure 1.7a shows that heat flux increased in general at all locations (note the difference in the scales of LHS and RHS y-axes) between the two extreme k_{wall} cases due to lower thermal resistance (t_{wall}/k_{wall}) of the wall. Only the inner tube contributed to heat recirculation in k_{wall} = 0.01 W/mK and the outer tube actually extracted heat from the reactant mixture in the preheating region (left of anchor point). The situation improved for the k_{wall} = 325 W/mK case and it is observed that both the tubes contributed to heat recirculation. Hence, inner tube contribution is critically important for lower wall conductivity and imparts additional stability at higher values of k_{wall}. Sharp change in the heat flux profile of 0.01 W/mK case indicates approximate location of the flame at which a sudden change in the direction of heat flux occurs. The transition is less abrupt for the 325 W/mK case and highlights the localization of heat reflux for low k_{wall}. This would lead to a less stable flame. Norton and Vlachos [22] also reported a similar observation. More heat recirculated at the higher inlet velocity (Figure 1.7b) in comparison to the lower velocity case and both inner and outer tubes contributed to heat recirculation. In this case, the heat transfer resistance on the fluid side ($1/h$) decreased to enhance the heat flux. Heat transfer from hot products to the outer wall is in general higher than to the inner wall in cases in Figure 1.7 since its thermal inertia and volumetric heat capacity are comparatively higher than the inner tube and N_2 core.

Gas bulk mean temperature and wall temperatures are compared in Figure 1.8 to show the relative importance of inner and outer tubes in preheating and thermal

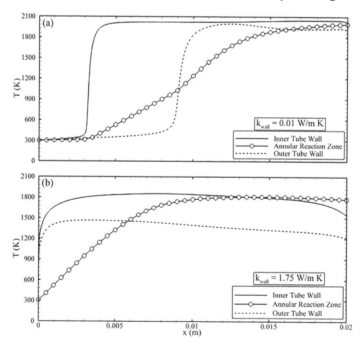

FIGURE 1.8   Longitudinal profiles of temperature on the inner and tube wall surfaces in contact with annular reaction zone compared with bulk mean temperature of the annular reaction zone; (a) k_{wall} = 0.01 W/mK, (b) k_{wall} = 1.75 W/mK. (Operating conditions: V_{in} = 15 m/s, T_{in} = 300 K, ϕ = 1.0.)

recycle. Heat transfer occurs from the inner tube wall to the annular reaction zone for low as well as high thermal conductivity walls. The outer tube takes part in thermal recycle if the wall thermal conductivity is high.

1.6.3 Flame Location and Shape

Location of flame indicates its stability at a given operating condition. Flame shape and location reveal the balance of transport processes in the combustor and are important components of the overall flame structure. Reaction rate and temperature fields are chosen as markers of flame location.

Figure 1.9 depicts contours of the reaction rate, indicating that the flame zone was composed of two distinct branches anchored to the outer and inner walls. The upper branch of the flame was shorter in comparison to the lower and both the branches extended axially along the flow direction. Higher wall heat losses at the outer tube–reaction zone interface resulted in a shorter upper branch, as seen from the data on total surface heat flux in Figure 1.7. Additionally, a zone of reduced reaction rate flanked the two branches. The backward flame movement in the colder region of the reactor accompanied by the higher inner tube temperature contributed to the asymmetry in the flame shape since the lower branch moved backward earlier than the upper one as observed from instantaneous contours of reaction rates in Figures 1.9a and 9b. Higher levels of temperature on the side of the inner tube induced the differential rate of

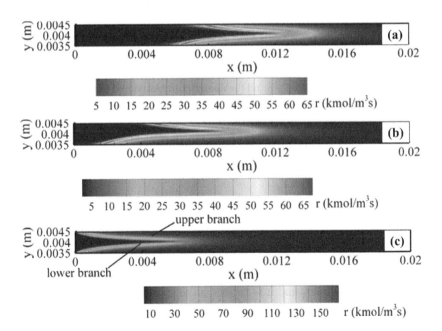

FIGURE 1.9 Contours of reaction rate (kmol/m^3 s) showing flame development; (a) $t = 2$ s, (b) $t = 8$ s, (c) steady state. (Operating conditions: $V_{in} = 15$ m/s, $T_{in} = 300$ K, $\phi = 1.0$, $k_{wall} = 1.75$ W/m K.)

movement. The outer tube wall lost more heat than the inner wall as suggested by low temperature levels, and this disparity led to dissimilar temperature distributions in the post-flame region. Thus, change in density in the post-flame region was not uniform along the radial direction, which also contributed to the asymmetry. The reaction rate reduced in the region flanking the two flame branches, which could indicate the flame being stretched under non-adiabatic conditions. Moreover, the wall temperature field governed the flame shape and its location in the microcombustor as explained earlier.

The outer wall absorbed the heat released during combustion and its temperature increased beyond the autoignition temperature for the reaction system resulting in the formation of an upper flame branch. Additionally, the outer wall also helped in anchoring the flame branch. A similar situation prevailed in the inner tube containing N_2, leading to the establishment of the lower flame branch. Low thermal conductivity of N_2 arrested the heat losses and transported the heat upstream to the cold reactant mixture. As a result, inner wall temperature also increased above the autoignition point of the fuel and thermal conditions on the inner tube–annular zone interface became more favourable as compared to the outer tube–annular reaction zone interface. Hence, the lower flame branch was longer than the upper and anchored closer to the inlet, extending beyond the centre plane of the reaction zone. Temperature variation across the flame showed sharp changes in the regions where the two flame branches were located. The presence of distinct flame branches and a pronounced asymmetry of the overall flame interface during the warm-up period of microcombustor can be observed from Figures 1.9a and 1.9b in comparison to the steady state shown in Figure 1.9c. As the heat content of the combustor increased and preheating progressed, both the flame branches moved backward towards the inlet and stabilized some distance away from the inlet. The resulting broadening of the reaction zone caused the flame branches to become progressively indistinct. The sharp decrease in temperature around the centre plane indicated that the flame shape was superficially similar to a typical Bunsen burner flame, of course with asymmetry due to the different anchoring points on the walls.

The annular microcombustor burning hydrogen achieved stable combustion in spite of the heat losses at the walls. Moreover, a stable flame was established without any need for catalytic action, which would considerably simplify the microcombustor design and operation.

1.7 PARAMETRIC STUDIES

Effects of flow rates, preheating, wall thermal conductivity, and geometric parameters on the stability of stoichiometric hydrogen–air flame and thermal performance of the annular microcombustor are considered. Quantification of flame location is necessary for such studies and it is defined as the longitudinal position of maximum reaction rate at the tube walls. At the outset, it is pointed out that numerical boundary conditions imposed at the flow inlet and outlet are increasingly violated as the flame approaches near these boundaries. Hence, it is difficult to obtain precise stability limits in numerical studies. Consequently, results are indicative of trends in the variation of flame positions. Nevertheless, they help to obtain approximate limits on stability and certainly indicate the approach to stability limits under given conditions.

1.7.1 INLET FLOW RATES

Inlet velocity (V_{in}) was varied in the range 3–25 m/s and flame response to the change in the velocity was characterized in terms of flame location at the walls. Velocity is restricted to 25 m/s at the upper limit because the corresponding Reynolds number (Re) based on hydraulic diameter of the annular reaction zone is smaller than the critical value. Transition Re_{crit} for the annulus flow increases with the radius ratio $K = r_{in}/r_{out}$ and approaches 2600 for $K = 0.778$ of the present case. Beyond Re_{crit}, flow can be laminar or turbulent and the problem is that of predicting transition from laminar to turbulent flow. This problem cannot be handled easily at present and one can only hope that the small grid size (25 μm) used throughout the flow region of the domain is sufficient to resolve the length scales present at $V_{in} > 25$ m/s ($Re > 2200$).

Figure 1.10 shows the variations in flame location with respect to inlet velocities parameterized by wall thermal conductivity. It is observed that the locations of the flame's leading edge on the inner and outer tubes did not change significantly for $k_{wall} = 1.75$ W/mK. In response to high momentum of incoming gas mixture, the flame stabilized farther away from inlet in the bulk flow and the flow straining induced stretch in the flame. It should be noted that the flame locations for the smallest inlet velocities (3 and 5 m/s) are at the first node location 25 μm away from inlet

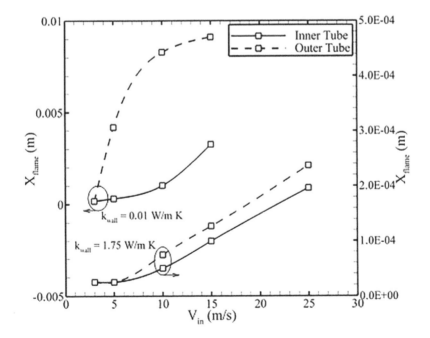

FIGURE 1.10   Variation in flame location with inlet velocity for stoichiometric feed.

and flame holding is most probably due to the frozen chemistry and low temperature at the inlet.

From the flame side, it can also be argued that while high temperatures of the surrounding structure sustained combustion, continuous removal of heat lost by the flame to the inlet (by means of a coolant or controlled temperature gradient) anchored flame near the inlet. Wall thermal conductivity being high, within the limits of single step kinetics, it can be argued that smoothing out of temperature gradients caused the backward movement of the flame along the walls of the inner and outer tubes up to the left end which acts as a heat sink. Reduced flow straining at the smaller velocities also reduced the stretch experienced by the flame and a flatter flame was obtained. While this reasoning holds good at high wall thermal conductivity, the limiting thermal conductivity for which this becomes possible cannot be reliably predicted with the use of a single step kinetics model.

Results shown in Figure 1.10 for k_{wall} = 1.75 W/mK indicate that edge flames will likely still be present at much higher velocities while flame in the bulk flow may be swept away by the high momentum flow. A flame as such might approach extinction in the bulk flow at high enough inlet velocities. At low velocities, an increasingly flat flame indicates a potential flashback.

For same inlet velocity and wall thickness, higher temperature gradients develop in the low thermal conductivity wall (roughly ten times larger as thermal resistance is inversely proportional to k_{wall}). This is detrimental to preheating of the reacting mixture and thermal coupling between the flame and wall is weakened. Consequently, the flame moves as a whole in response to velocity changes. This reasoning also explains the upstream location of the flame for lower inlet velocities. Resistance to flame movement upstream is also present in the flow by virtue of a local velocity component normal to the flame front, and thermal resistance in the high wall thermal conductivity case is weak enough to permit flame movement along the walls for a range of velocities. In case of low thermal conductivity, the relative dominance of the two resistances (thermal in walls, kinematic in fluid) decides flame location.

Flame stability limits can be recognized definitively at low k_{wall}. The bulk movement of flame towards an outlet indicates a blowout-like situation at higher velocities. Note that maximum wall temperatures are far higher than the autoignition point at the highest velocity considered here for k_{wall} = 0.01 W/mK so the flame will still ignite. The upstream movement in response to decrease of inlet velocities has already been explained and it is noted that the flame approaches flashback at smaller velocities for more insulating wall materials.

For a given k_{wall}, any shift in the leading edge location in response to change in the inlet velocity also depends upon where the flame stabilizes in the bulk flow. If the flame successfully opposes the incoming gases, then consequent heating of the wall will lead to an increase in heat release rate near the wall and a strong flame leading edge will establish at the wall. Subsequent upstream movement of the flame leading edge and its steady state location is as explained previously.

Overall effects of velocity changes are shown in Figure 1.11 in terms of the volume average temperatures in the four principal zones. Temperature data for low thermal conductivity (k_{wall} = 0.01 W/mK) in Figure 1.11a show that average temperatures in all zones dropped at higher velocities due to bulk movement of the

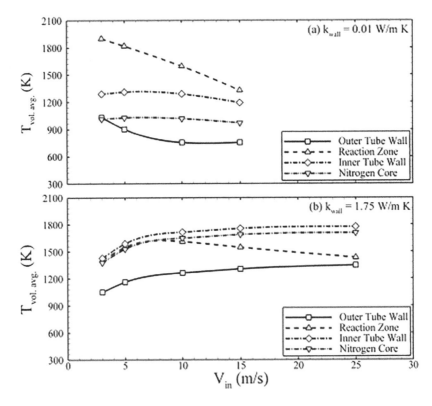

FIGURE 1.11 Average zonal temperatures as a function of inlet velocity (a) k_{wall} = 0.01 W/mK, (b) k_{wall} = 1.75 W/mK.

flame towards blowout. High average temperatures in the walls at lower velocities (before the flashback limit) indicate that low velocity operation is better for insulating materials since chances of thermal quenching are low. The trends however almost entirely reversed in case of higher thermal conductivity (k_{wall} = 1.75 W/mK) as seen from Figure 1.11b and the reaction zone cooled considerably at velocities >10 m/s. This is because the flame elongated and aligned itself with the flow at higher velocities, which increased the fraction of microcombustor volume upstream of the flame occupied by cold mixture. Flame elongation due to flow straining is more in this case because low thermal resistance of high conductivity walls maintained high temperatures and kept up preheating.

Results reported here for stoichiometric flame indicate that kinetics probably plays a minor role and thermal conditions in the wall determine the flame behaviour. Secondly, power output of the microcombustor also increased at higher velocities. Consider the issue of lower flame stability limit for highly conductive walls alluded to previously. If walls cool down at smaller velocities, then it is likely that the flame will suffer global extinction instead of flashback as illustrated by Figure 1.11b. Although average temperatures of inner and outer tube (1440 K and 1020 K) are much higher

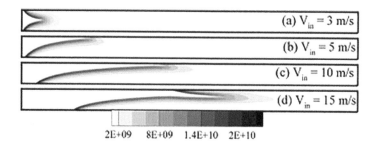

FIGURE 1.12 Heat release rate profiles (in W/m³) for k_{wall} = 0.01 W/mK and different inlet velocities (V_{in}): (a) 3 m/s, (b) 5 m/s, (c) 10 m/s, (d) 15 m/s.

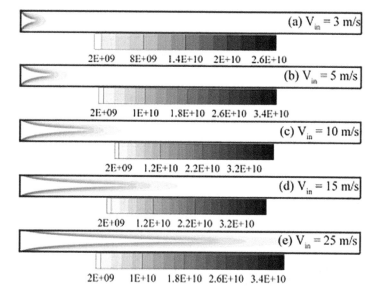

FIGURE 1.13 Heat release rate profiles (in W/m³) for k_{wall} = 1.75 W/mK and different inlet velocities (V_{in}): (a) 3 m/s, (b) 5 m/s, (c) 10 m/s, (d) 15 m/s, (e) 25 m/s.

than the autoignition temperature for hydrogen (783 K), it is conceivable that for longer tubes, they will further decrease and might cause global extinction.

Approach to flashback is clearly observed for 0.01 W/mK but the data on 1.75 W/ mK are less conclusive. The flame will approach flashback at low enough velocities in the case of an annular microcombustor (as against extinction by thermal quenching) since hydrogen has a higher burning velocity than methane and thermal inertia of the inner and outer tubes supply enough heat to self-sustain combustion at low flow rates.

Overall flame response to velocity increase is shown in Figures 1.12 and 1.13 by volumetric heat release rate in W/m³. Figure 1.12 is for k_{wall} = 0.01 W/mK in the velocity range of 3–15 m/s and Figure 1.13 shows data for 1.75 W/mK in the velocity range of 3–25 m/s. A distinct feature of flame in the case of 0.01 W/mK is the pronounced asymmetry in flame shape at higher velocity while flame asymmetry decreased at $k_{wall} = 1.75$ W/mK. Reasons for this asymmetry lie in the structural features of annular microcombustor that favour lower flame branch as discussed now. Higher inlet velocities caused flame to establish away from the inlet and outer tube wall temperature remained always lower than the inner tube. On the other hand, thermal isolation of the inner tube retained the heat supplied by flame, which kept the post-flame gases at high temperatures. Consequently, more and more hydrogen burnt and the flame branch near the inner tube elongated. Otherwise, hydrogen would have escaped unreacted in the absence of high post-flame temperatures, had the heat losses from the inner tube been high.

Performance of the annular microcombustor is analyzed to determine its suitability for electric power generation and micro propulsion applications using non-dimensionalized terms in the heat balance, pattern factor, and efficiencies. Their definitions are to follow. In order to explain the effect of higher inlet temperature on heat balance, heat reflux ratio (Q_{HR}) and heat loss ratio (Q_{HL}) are defined in eq. (3) and (4), respectively.

$$Q_{HR} = \frac{\dot{Q}_{reflux}}{\dot{Q}_{wall}} \times 100 \tag{3}$$

$$Q_{HL} = \frac{\dot{Q}_{wall}}{\dot{m}_f LHV} \times 100 \tag{4}$$

where Q_{HR} is the ratio of rates of heat reflux in the preheat region ($\dot{Q}_{reflux}$) to the wall heat losses ($\dot{Q}_{wall}$). Q_{HL} is the ratio of heat loss rate to combustor thermal power. Heat reflux rate was computed in the preheat region as the actual heat transferred to the reactants from walls. Another relevant indicator of performance defined by eq. (5) is the thermal output ratio ($Q_{O/P}$), which compares heat transfer rate at the outlet ($\dot{Q}_{outlet}$) to the wall heat losses.

$$Q_{O/P} = \frac{\dot{Q}_{outlet}}{\dot{Q}_{wall}} \times 100 \tag{5}$$

Suitability of the annular microcombustor for gas turbine–based application as well as TPV or TE energy conversion depends additionally upon the uniformity of temperature profiles at outlet and longitudinal walls, respectively. Uniform temperature profiles are generally preferable. This possibility is judged by the non-uniformity of temperature at outlet as quantified by the pattern factor (PF) of eq. (6).

$$PF = \frac{\left(T_{max} - T_{avg}\right)}{\left(T_{avg} - T_{in}\right)}\Bigg|_{outlet\ OR\ wall} \tag{6}$$

T_{avg} and T_{max} denote the maximum and average temperature values at the outlet and wall, depending upon the locations for evaluation of pattern factor. Combustion efficiency (η_{comb}), thermal efficiency ($\eta_{thermal}$), and overall efficiency ($\eta_{overall}$) relating to the thermal performance of combustor in propulsion application are defined next in terms of the enthalpies at outlet, wall, and inlet.

$$\eta_{comb} = \frac{\dot{Q}_{outlet} - \dot{Q}_{inlet} + \dot{Q}_{wall}}{\dot{m}_f LHV} \times 100 \tag{7}$$

$$\eta_{thermal} = \frac{\dot{Q}_{outlet} - \dot{Q}_{inlet}}{\dot{Q}_{outlet} - \dot{Q}_{inlet} + \dot{Q}_{wall}} \times 100 \tag{8}$$

$$\eta_{overall} = \frac{\dot{Q}_{outlet} - \dot{Q}_{inlet}}{\dot{m}_f LHV} \times 100 \tag{9}$$

Additionally, the completeness of combustion is defined by the conversion efficiency in terms of H_2 mass fraction (Y_{H_2}), which is the limiting species of the mixture.

$$X_{H_2} = \left(1 - \frac{Y_{H_2}\big|_{out}}{Y_{H_2}\big|_{in}}\right) \times 100 \tag{10}$$

The chemical enthalpy input increased at the inlet as inlet mass flow rates increased, causing proportionate rise in the levels of heat transfer rates. Figure 1.14a shows

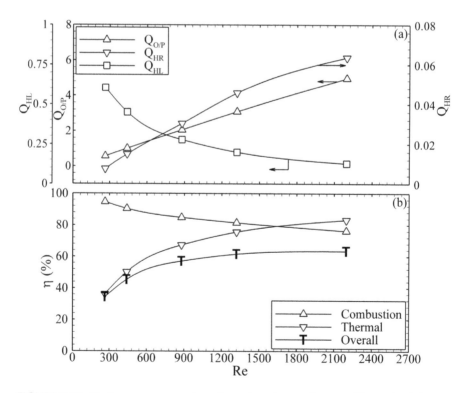

FIGURE 1.14 Performance parameters of annular microcombustor with varying inlet *Re*. (a) heat transfer ratios, (b) efficiencies (k_{wall} = 1.75 W/mK, V_{in} = 3–25 m/s).

the effects of increasing mass flow rates on the thermal ratios in terms of Reynolds number (*Re*) at the inlet defined using the hydraulic diameter of annular reaction zone (D_h) as the characteristic length scale. Although heat losses at the walls were higher at higher mass flow rates, the increased heat reflux rate and enthalpy transfer rate at the outlet compensated for the effects of increasing heat losses at the walls. Consequently, both the heat reflux ratio (Q_{HR}) and thermal output ratio ($Q_{O/P}$) increased at higher flow rates. A higher heat reflux ratio is beneficial for the flame stabilization. $Q_{O/P} > 1.0$ means that combustor would be more suitable for propulsive application, while $Q_{O/P} < 1.0$ indicates that it will be more suitable for heating and electric power generation by thermoelectric or thermophotovoltaic routes. Higher $Q_{O/P}$ would result in better performance for the micro propulsion application since more heat would be channelized to the outlet in comparison to the wall losses. A positive aspect of operating at the higher mass flow rates is also revealed in the diminished heat loss ratio (Q_{HL}) in Figure 1.14. However, the decrease in Q_{HL} is marginal in this case.

The net thermal output enhanced with increased input of the chemical enthalpy at higher velocities. However, the microcombustor stored part of the heat, which reduced the thermal output in relation to the chemical enthalpy input and resulted in the decreased combustion efficiency. On the other hand, the drastic increase in availability of enthalpy at the outlet offset the effects of heat losses so that the thermal efficiency increased as inlet mass flow rates increased. Thus, data on the thermal performance of a microcombustor in Figure 1.14b revealed that operation at higher mass flow rates improved thermal efficiency but reduced the combustion efficiency. The overall performance at different inlet velocities is determined by the thermal efficiency and the overall efficiency increased with the mass flow rates, eventually reaching a plateau beyond the inlet mass flow rate corresponding to 15 m/s as seen from Figure 1.14b.

1.7.2 INLET TEMPERATURE

Higher inlet temperature of reactant mixture simulates external preheating, which could be useful for a less energetic fuel. However, it is important to understand the effect of inlet temperature (T_{in}) on the overall flame structure.

Contours of heat release rate in Figure 1.15 showed dramatic changes in response to increase in T_{in} from a slender flame at low T_{in} to a flatter one at higher temperatures. The heat release rate increased near the mid-plane due to reduction in the preheating zone and uniform ignition across the channel. Calculations done with lower heat losses predicted more compact flame with less diffuse heat release zone at all temperatures. Representative changes in the flame temperature and reaction rate are shown at the mid-plane of annular reaction zone in Figure 1.16a for the three inlet temperatures.

The reaction rate increased and reaction zone narrowed at higher T_{in} due to enhanced burning rates, while the flame temperature increased proportionately. However, heat losses in the post-flame regions, as evident from a continuous decrease in temperature profiles, reduced the outlet temperatures for higher T_{in}, even below the base case. These trends point towards a further drop in temperature for a higher

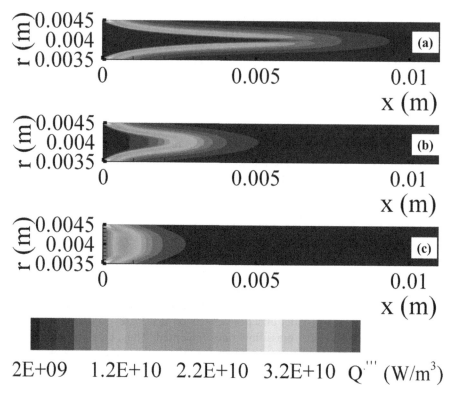

FIGURE 1.15 Contours of heat release rate (a) T_{in} = 300 K, (b) T_{in} = 400 K, (c) T_{in} = 500 K.

aspect ratio combustor. They are controlled by two factors; although higher T_{in} led to flame establishment near the inlet, heat losses from the region near the inlet to ambient increased due to the enhanced temperature gradient. Secondly, axial heat conduction in the wall transferred the heat away from flame. In this respect, preheating may not provide much advantage.

However, the inner tube could prove advantageous as inferred from the axial profiles of temperature at the outer surfaces of the tubes in Figure 1.16b. It is observed that the preheating marginally raised the outer wall temperature near the inlet region (~50 K for a 200 K rise in T_{in}), limited by the heat losses to ambient. However, the inner tube temperature for most of its length was always higher at any T_{in} due to thermal isolation of its walls by the flame on the outer side and (low thermal conductivity) N_2 on the inner side. Although the rise in wall temperature in comparison to the rise in inlet temperatures was marginal on the inner tube side too, the heat available in the reactant mixture propagated downstream predominantly due to axial heat transfer in the walls. Lower thermal conductivity of N_2 in the immediate vicinity of the inner tube wall offered higher resistance to heat conduction in the radial direction and thereby enhanced axial heat transfer. These results indicate that the inner

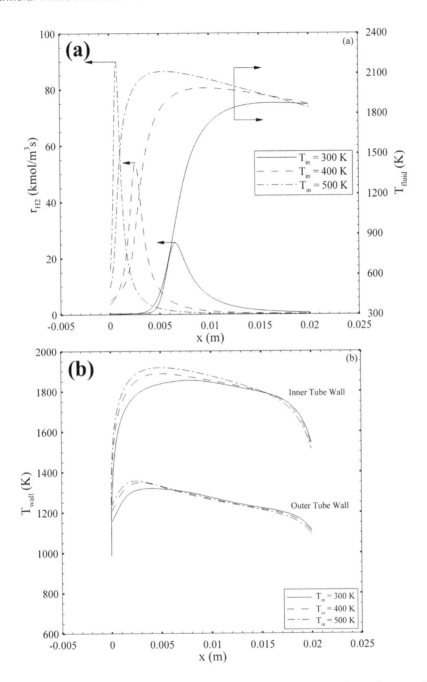

FIGURE 1.16 Effect of inlet temperature on (a) flame temperature and reaction rate, (b) temperature profiles.

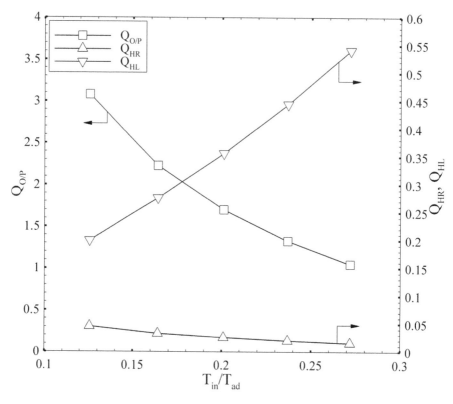

FIGURE 1.17   Effect of inlet temperature on the heat balance of annular microcombustor.

tube would be more effective at higher T_{in}, but the final choice of inlet temperatures would also be dictated by the overall thermal performance.

Figure 1.17 shows the dependence of thermal ratios on inlet temperature normalized by respective T_{ad}. Q_{HR} decreased drastically with increase in T_{in} (66% in comparison to the base case). It occurred due to a reduction in the preheating length at higher T_{in} and a small amount of recirculated heat from the hot products was sufficient to raise mixture temperature to ignition temperature and establish flame near the inlet, as also seen in Figure 1.16a. It is also observed from the data for $T_{in} = 300$ K that the combustor could reflux a significant amount of heat through the walls. On the other hand, overall heat losses increased drastically as suggested by increased Q_{HL}, reducing the available heat which otherwise could be transferred for the preheating of incoming fuel-air mixture.

It will be noted that the present configuration has lower effective surface area per unit volume, which aids in heat conservation. Hence, reduced Q_{HR} and increased Q_{HL} make higher T_{in} operation less desirable in comparison to the lower T_{in}. This analysis shows that it may not be necessary to use preheated fuel–air mixture to establish a self-sustained flame in the proposed microcombustor, which can simultaneously prevent material failure and limit the pollutant (NO_x) emissions. Further, preheating is inherently unsafe for premixed combustion and should be kept to a minimum level. Secondly, higher temperature gradients established with the ambient in case of preheated mixtures led to higher downstream heat losses and non-isothermal walls. Higher heat losses and the diminished heat reflux rates at the higher T_{in} could impede flame stability and higher inlet temperatures might narrow down the stable operating range of a high-energy fuel like hydrogen by an early flashback.

Apart from inlet temperature, $Q_{O/P}$ also depends upon wall thermal conductivity, thermal capacity of solid, type of fuel, inlet velocity, etc., all of which are kept constant in this study. Figure 1.17 shows that $Q_{O/P} > 1.0$ for comparatively lower T_{in} cases up to 500 K and it decreased by 66% at the highest T_{in}. At 700 K, $Q_{O/P} \oplus 1$, which indicates that the annular combustor could be suitable for propulsion application at lower inlet temperatures and as a thermal source for TPV or TE energy converters at medium to high values of inlet temperatures since the availability of heat at the walls increased at higher T_{in}.

Values of PF for outlet (open symbols) in Figure 1.18a showed an increase of about 46% with inlet temperature and a steep increase in PF (~58%) is also predicted for the external longitudinal wall (filled symbols). PF at the outlet varied between 0.12 and 0.19, which compares favourably with the recommended range of 0.15–0.25 for high temperature–rise gas turbine combustors. Hence, the annular microcombustor could be useful for gas turbine propulsion at the inlet temperature adjusted for material failure, safety, and fuel energy content. Similarly, the pattern factor for wall was also defined. Although, operation at higher T_{in} raised the availability of heat at the external wall surface as shown by the increased Q_{HL}, it also resulted in increasingly non-uniform wall temperatures, so that the conversion efficiency of a thermoelectric or thermophotovoltaic generator may be adversely affected. The resulting increase in the pattern factor for longitudinal wall was much more substantial than the outlet.

Thus, an annular microcombustor should be operated without much external preheating of the fuel-air mixture. Thermal recycle from hot products can fulfil this requirement. Operation at lower inlet temperature maximizes its thermal performance in terms of higher $Q_{O/P}$, higher Q_{HR}, and lower Q_{HL}. Operation at lower T_{in} would also provide uniform temperature profiles at the external wall and outlet for electric power generation and propulsion applications without additional conditioning of the exhaust gases.

Figure 1.18b shows the effect of inlet temperature on various efficiencies and conversion of limiting reactant. H_2 conversion efficiency is high at all temperatures with only a slight increase at higher temperatures. Results of multi-step kinetics calculations for the same conditions indicate that single step kinetics predicted conversions (96–98%) are marginally higher than multi-step data (94–96%). Combustion efficiency actually increased with T_{in} as higher temperatures in the radial direction presented more locations for burning hydrogen. The flame became more compact and could be sustained in the central region at higher inlet temperatures. Localized weakening of the flame in the central regions reduced its heat release rate at low T_{in}, causing lower combustion efficiency. Total thermal output of the microcombustor (at the outlet and walls) increased at higher inlet temperatures, which increased the combustion efficiency. On the other hand, thermal efficiency decreased since the heat losses increased and thermal output at the outlet decreased in proportion. Thermal

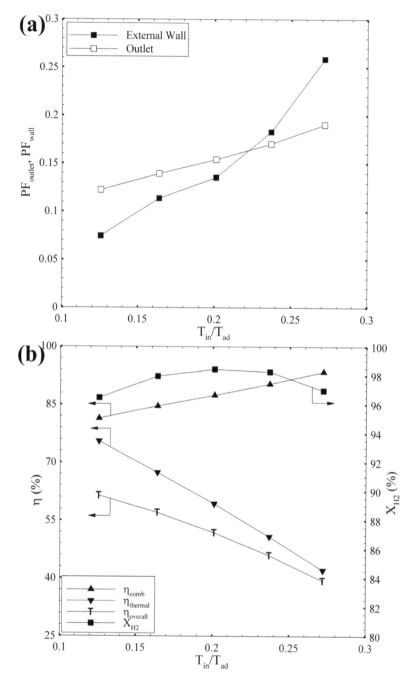

FIGURE 1.18 Variation in thermal performance with inlet temperature (a) pattern factor, (b) efficiencies.

output ratio ($Q_{O/P}$) verified this observation. The disparity in the proportion of heat losses at the walls and outlet enthalpy led to a reduction of the exhaust temperatures as inlet temperatures increased. These results indicate that the combustor will operate more efficiently at lower inlet temperatures, as overall efficiencies are high. The overall efficiency value at low temperatures is in the range 51–61% and compares favourably with the reported value [23] for stoichiometric H_2-air combustion.

1.7.3 CHOICE OF OPERATING CONDITIONS

Leach and Cadou [24] noted that the maximization of thrust (power) and range (endurance) might not be achieved in the same microcombustor design. This aspect was investigated here with respect to the different inlet flow rates and wall thermal conductivities. Power density limits the thrust while overall efficiency dictates the achievable range in a propulsion device.

Figure 1.19 shows the data for these parameters at different mass flow rates and thermal conductivities discussed earlier. It was found that the operating conditions for maximum efficiency and maximum power density were different for the same

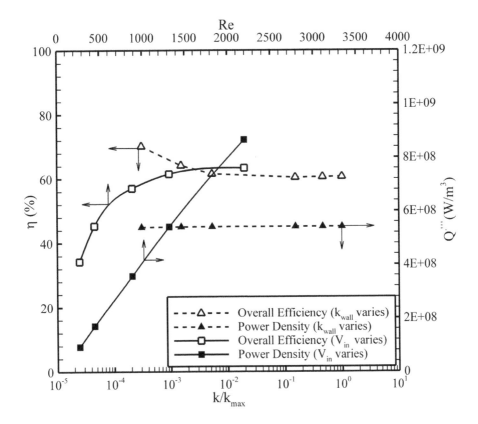

FIGURE 1.19 Thrust-range relationships for different parameters.

annular microcombustor design. In particular, maximum power density operating point was obtained at a higher mass flow rate approaching the blowout value in comparison to the high overall efficiency point since the power density is directly proportional to the mass flow rate. Power density increased only marginally in the case of different thermal conductivities.

1.8 SUMMARY

Results of the numerical analysis of the novel annular microcombustor indicate that the inner tube and nitrogen gas stored inside acted as a source of heat during the heating phase of the combustor and stabilized the flame on the walls of the combustor tubes without any flame holder. Asymmetric flame and detailed modelling of heat and mass transfer processes are the two important aspects of this study. Heat losses at the walls caused non-uniform temperature distributions in the post-flame zone that gave an asymmetric shape to the flame.

In order to fulfil the main goal of realizing a microcombustor for propulsive applications, an annular microcombustor relying solely on internal heat recirculation and flame holding by hot walls was devised. It possesses additional features of reduced solid structure, simpler and compact design, and low pressure drop. Performance characterization of the annular microcombustor provided guidelines for efficient operation. Proposed design is useful as a source of propulsive energy and thermoelectric or thermophotovoltaic power with proper choice of wall material and adjustments in the operating conditions. This versatility comes from a combination of shape factor and ability to stabilize the flame near inlet without the need for structural or fluid dynamic flameholders.

REFERENCES

1. B. Lewis and G. von Elbe. *Combustion Flames and Explosion of Gases.* Academic Press, Cambridge, MA, 1961.
2. R.B. Peterson. Miniature and microscale energy systems. In M. Faghri and B. Sunden (Eds.). *Heat and Fluid Flow in Microscale and Nanoscale Structures*, WIT Press, London, 2003.
3. I.A. Waitz, G. Gauba, and Y.-S. Tzeng. Combustors for micro-gas turbine engines. Trans. *ASME J. Fluids Eng.* 120 (1998), pp. 109–117.
4. A.H. Epstein. Millimeter-scale, micro-electro-mechanical systems gas turbine engines. *J. Eng. Gas Turbine Power* 126 (2004), pp. 205–226.
5. C.M. Reid, M.A. Manzo, and M.J. Logan. Performance characterization of a Li-ion gel polymer battery power supply system for an unmanned aerial vehicle. NASA/TM-2004–213401, 2004–01-3166 (2004), pp. 1–13.
6. S.-B. Lee, T.-H. Lim, R.-H. Song, D.-R. Shin, and S.-K. Dong. Development of a 700 W anode-supported micro-tubular SOFC stack for APU applications. *Int. J. Hydrogen Energy* 33 (2008), pp. 2330–2336.
7. D.G. Norton, E.D. Wetzel, and D.G. Vlachos. Catalytic microcombustors for compact power or heat generation. US20070082310A1 (2007).
8. D. Dunn-Rankin, E.M. Leal, and D.C. Walther. Personal power systems. *Prog. Energy Combust. Sci.* 31 (2005), pp. 422–465.

9. A.H. Epstein, S.D. Senturia, I.A. Waitz, J.H. Lang, S.A. Jacobson, F.F. Ehrich, M.A. Schmidt, G.K. Ananthasuresh, M.S. Spearing, K.S. Breuer, and S.F. Nagle. Microturbomachinery. US20026392313B1 (2002).

10. D.W. Youngner. MEMS microthruster array. US20026378292B1 (2002).

11. A.L. Cohen, P.D. Ronney, U. Frodis, L. Sitzki, E.H. Meiberg, and S. Wussow. Microcombustor and combustion based thermoelectric generator. US 6,951,456 B2 (4 October 2005).

12. S.K. Chou, C. Shu, W.M. Yang, H. Xue, and Z.W. Li. Thermophotovoltaic power supply. US 7,557,293 B2 (7 July 2009).

13. G.F. Carrier, F.E. Fendel, and P.S. Feldman. Laminar flame propagation/quench for a parallel-wall duct. *Symp. (Int.) on Combustion* 22 (1967), pp. 67–74.

14. E.W. Lemmon. Thermophysical properties of fluids. In D.R. Lide and W.M. Haynes (Eds.). *CRC Handbook of Chemistry and Physics, 90th Edition 2009–2010, Section 6: Fluid Properties*, 6.21–6.37, CRC Press, Boca Raton, FL, USA, 2009.

15. *Fluent 6.3 User's Guide*, Fluent Inc., NH, USA, 2006.

16. Y.S. Touloukian and D.P. DeWitt. *Thermal Radiative Properties: Metallic Elements and Alloys*, Thermophysical Properties of Matter, The TPRC Series, 1970, Defence Technical Information Centre, Vol. 7, Fort Belvoir, VA.

17. S.Y. Jejurkar. Numerical studies on hydrogen-air premixed flame based annular microcombustor. PhD Thesis, IIT Kanpur, 2011.

18. S.Y. Jejurkar and D.P. Mishra. Some aspects of stabilization and structure of laminar premixed hydrogen-air flames in a microchannel. *Appl. Therm. Eng.* 87 (2015), pp. 539–546.

19. K. Hsu and A. Jemcov. Numerical investigation of detonation in premixed hydrogen-air mixture—assessment of simplified chemical mechanisms. AIAA–2000–2478, AIAA, Fluids 2000 Conference and Exhibit, Denver, CO (2000).

20. J. Li, S.K. Chou, Z.W. Li, and W.M. Yang. Characterization of wall temperature and radiation power through cylindrical dump micro-combustors. Combust. *Flame* 156 (2009), pp. 1587–1593.

21. S.T. Lee and J.S. T'ien. A numerical analysis of flame flashback in a premixed laminar system. *Combust. Flame* 48 (1982), pp. 273–285.

22. D.G. Norton and D.G. Vlachos. Combustion characteristics and flame stability at the microscale: A CFD study of premixed methane/air mixtures. *Chem. Eng. Sci.* 58 (2003), pp. 4871–4882.

23. D.B. Spalding, H.C. Hottel, A.H. Lefebvre, D.G. Shepherd, and A.C. Scurlock. The art of partial modeling. *Symp. (Int.) on Combustion* 9 (1963), pp. 833–843.

24. T.T. Leach and C.P. Cadou. The role of structural heat exchange and heat loss in the design of efficient silicon micro-combustors. *Proc. Combust. Inst.* 30 (2005), pp. 2437–2444.

2 Inverse Jet Flame Based Swirl Combustor

Mahesh S. and Debi Prasad Mishra

CONTENTS

2.1 GAS TURBINE COMBUSTOR

A combustor, in the context of gas turbine engines, is a sub-entity/region within the engine where the mixing of fuel with oxidizer leading to an exothermic chemical reaction takes place. A reliable combustor should be capable of anchoring a stable flame established from the oxidation of reactants and ensure complete combustion within a compact volume. Importantly, a combustor should help in establishing a near uniform temperature distribution at the turbine entry for preventing hot spots and uneven thermal load on rotating turbine blades. In addition, reliable and hassle-free ignition of the reactants (fuel-air) during engine start-up and subsequent relighting after unexpected flame blowout in high altitudes are the major requirements expected of a combustor. Apart from this, enhanced flame stability limit over a wide range of air-fuel ratio within the combustor along with fuel flexibility are other essential requirements sought. The combustor should also incur low pressure loss and ensure low pollutant emissions such as NO_x, CO and soot in the exhaust through optimum air-fuel ratio.

The combustor geometry, its size and shape is predominantly dictated by the engine requirements. A typical gas turbine combustor of an aircraft engine consists of an outer air casing duct and an inner flame tube with liners. The air exiting the compressor initially enters the diffuser section of the combustor where a portion of it is channelled to the flame tube and the remaining part of it is diverted to the annular region between the outer casing and the flame tube. The diffuser at the entrance of the combustor helps in the smooth deceleration of air velocity thereby reducing

DOI: 10.1201/9781003049005-2

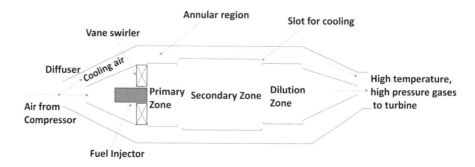

FIGURE 2.1   Schematic of various features in a gas turbine combustor.

the pressure loss. The air diverted to the annular region of the combustor is mainly utilized for cooling the flame tube which anchors the flame and also for diluting the hot gases exiting the combustor. The dilution of the combustion exhaust with the air passing through the annular region helps in ensuring complete combustion and also maintains a near uniform temperature distribution at the combustor exit plane for alleviating the thermal stresses on the rotating turbine blades.

The schematic of a typical gas turbine combustor is shown in Figure 2.1. The region within the combustor can be broadly divided into three zones—namely, (i) primary zone, (ii) secondary or intermediate zone and (iii) dilution zone. The role of the primary zone is to help in the mixing of fuel and air that enters the flame tube and anchors a stable flame. The upstream of the primary zone houses the fuel injector for atomizing the liquid fuel which is essential for its faster evaporation and mixing with the swirl air surrounding it. The fuel injector is enveloped by the vane swirler for imparting tangential motion to the air entering the primary zone. The swirl motion imparted to the air helps in creating a stable toroidal recircula-tion region in the primary zone for stabilizing a stable flame. The toroidal recircu-lation zone is a low pressure region which anchors the flame by recirculating hot product gases and excited radicals for continuous ignition of the incoming fresh reactants. The temperature within the primary zone will be in the order of 2000 K which can lead to dissociation reactions leading to higher levels of CO and H_2 in the hot products [1]. The intermediate zone of the combustor dilutes the hot products coming from the primary zone with the annular air for enhancing the oxidation of CO, unburned hydrocarbons and soot particles in the hot products. Subsequently, the dilution zone helps in further dilution of the hot products with the annular air for achieving a near uniform temperature of the product gases (refer to Figure 2.1) before it encounters the turbine section. Twenty to forty percent of the total combus-tor airflow is dedicated for this purpose in the dilution zone.

2.2   NONPREMIXED FLAME

Combustion process in a gas turbine combustor involves rapid oxidation of the fuel in a thin zone termed as 'flame'. Flame can be conceived as a thin spatial zone where rapid exothermic reactions take place with the emission of heat and light [2]. An important characteristic of the flame that distinguishes it from other chemical reactions is its ability to self-sustain by propagating into the combustible mixture with a subsonic velocity. The flame can be broadly classified into (i) premixed flame and (ii) nonpremixed flame based on the state of mixing between the fuel and oxidizer prior to combustion. In the case of premixed flame, fuel and oxidizer are mixed to the molecular level before the initiation of combustion. The type of flame established with initially unmixed fuel and oxidizer is termed as nonpremixed flame. Nonpremixed flame provides safety and stability over a wide operating range as compared to the premixed flame and hence, is employed in aircraft and stationary gas turbine engines, rocket combustors, diesel engines, industrial furnaces, etc. The nonpremixed flame can be classified further into two types—namely, (i) free jet flame and (ii) coaxial jet flame—based on the mode of delivery of the reactants. The nonpremixed flame established by the fuel jet injected into a quiescent oxidizer environment in either swirl or non-swirl mode is termed as a free jet flame [3]. In the free jet flame, entrainment of the oxidizer from the quiescent atmosphere depends upon the exit velocity of the fuel jet which implies that the oxidizer flow cannot be controlled independently in this nonpremixed flame type. However, this problem can be overcome by using a coaxial jet flame configuration, where oxidizer and fuel jets are delivered separately through a coaxial tube nozzle/burner with or without induced swirl motion.

Coaxial jet flame configuration is employed in gas turbine engines, rocket combustors, industrial burners, oxy-fuel welding etc. because the global flame characteristics can be manipulated by controlling the momentum of fuel and oxidizer jets independently. The coaxial jet flame configuration can be further classified based on the air-fuel port arrangement used for air-fuel jet delivery. The nonpremixed flame established in a coaxial tube nozzle/burner with central fuel jet and annular air jet stream is termed as Normal Jet Flame (NJF) and the flame established with the central oxidizer jet and annular fuel jet in a coaxial tube nozzle/burner is referred as Inverse Jet Flame (IJF). The schematic of the port configurations for NJF and IJF configuration are shown in Figure 2.2.

2.3   LEAN PREMIXED COMBUSTION—ISSUES AND ALTERNATIVES

Extensive research in the last three decades has indicated that operating gas turbine combustors in the fuel lean regime (equivalence ratio $(\Phi) < 1$) with excess air results in the reduction of product gas temperature and subsequent minimization of NO, CO, HC and soot emissions. One method for ensuring lean combustion is to premix the fuel with excess air before it enters the primary zone and this mode of combustion is commonly referred to as lean premixed combustion. However, lean premixed combustion is susceptible to flame stability issues such as lean blowout, thermoacoustic instability and flame relighting failure in high altitude flight conditions. Even though premixed flame is a better candidate for reducing pollutant emissions such as soot and NO_x emissions, they can lead to a serious safety hazard

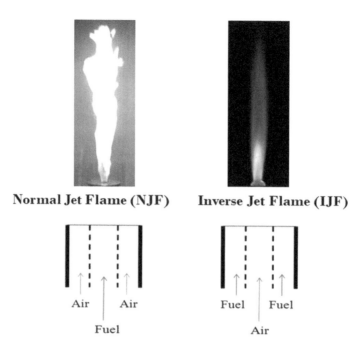

Normal Jet Flame (NJF)   **Inverse Jet Flame (IJF)**

FIGURE 2.2   Schematic of air-fuel port arrangement for NJF and IJF.

by propagating back (flashback) into the unburned premixed reactants under certain conditions when the flame speed becomes greater than the reactant flow speed.

From a safety perspective, one way to overcome the flashback issue in lean premixed combustion in gas turbine combustors is to stream the air jet and prevaporized fuel spray separately in a nonpremixed mode through a coaxial tube nozzle configuration. With respect to the coaxial jet flame, recent studies by researchers revealed that the coaxial jet flame established with a central oxidizer jet and annular fuel jet (i.e., the IJF) exhibits superior characteristics as compared to the normal jet flame. The high velocity central air jet augments the entrainment of the fuel jet in inverse jet flame configuration as compared to the NJF. Moreover, rapid fuel-air mixing can be achieved due to enhanced shear and turbulent intensity at the interface between two jets which results in a compact non-luminous flame with shorter flame height [4–10]. In addition to this, the flame shape, thermal and emission characteristics of inverse flame can be controlled and optimized with independent control of air and fuel jets [4–11]. Due to these favourable characteristics of IJF, it is expected that it can be a suitable alternative for lean premixed flame. Hence, an experimental study is undertaken to investigate the global characteristics of IJF and analyze the feasibility of employing this flame configuration in gas turbine combustors. In this chapter, the effect of the IJF characteristics—namely, visible flame appearance, visible flame height, lean blowout limits, exit temperature distribution and post flame NO_x emission levels—are studied experimentally by varying the swirl strength of the central air jet in a quartz tube confined dump combustor.

2.4   INVERSE JET FLAME BASED SWIRL COMBUSTOR
DEVELOPMENT AND EXPERIMENTAL METHODS

The inverse jet flame based swirl combustor discussed in this chapter is a quartz tube confined dump combustor which anchors a swirl inverse jet flame. The schematic of the swirl IJF combustor is shown in Figure 2.3. The combustor utilized in the present study consists of a coaxial burner with central swirl chamber for delivering the swirl air jet and an annular passage for injecting the fuel jet. The swirl chamber consists of four tangential ports each of 3 mm diameter (d_t) for injecting tangential air jets. Apart from the tangential air jets, air also enters axially into the swirl chamber. The swirl is imparted to the axial air with the help of discrete air jets injected through the tangential ports of the swirl chamber. In this study, axial air flow rate is maintained constant at 5 litres per minute and the swirl intensity of the central air jet is varied by gradually increasing the tangential air flow rate through the swirl chamber. Another distinct feature of the coaxial burner is that the swirl chamber exit is recessed by 9.3 mm within the annular passage for effective entrainment of the annular fuel jet by the central swirl air jet. The swirl IJF anchored in the coaxial burner is confined with a 7.96 cm diameter quartz tube of length 34.9 cm. The Compressed Natural Gas (CNG) is used as fuel for the present experimental study. The swirl number of the central air jet is calculated based on Eq. (1) proposed by Sautet et al. [12], which is given by

$$SN_a = 0.75\frac{\dot{m}_t}{\dot{m}_{ax}} \tag{1}$$

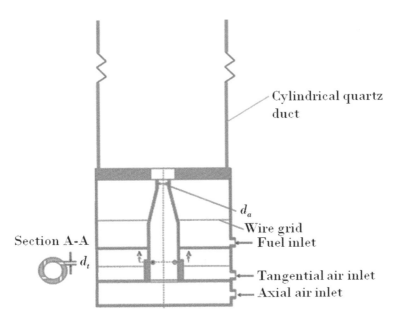

FIGURE 2.3   Schematic of the swirl IJF based combustor.

where m_{ax} and m_t denote the axial air and tangential air mass flow rates fed through the central tube of the coaxial burner. The swirl number calculated using Eq. (1) will be hereafter referred as aerodynamic swirl number, SN_a in this chapter.

The visible appearance of swirl IJF within the combustor at different annular fuel jet velocities and air jet aerodynamic swirl number cases is acquired at 30 frames per second using a SONY HDR-XR100 video camera. The temperature distribution at the combustor exit plane is measured with the help of a R-type (Pt/Pt-13%Rh) thermocouple having a bead diameter of 0.4 mm. The catalytic reaction on the thermocouple bead is prevented by coating it with alumina based ceramic called Ceramabond 569. The unsteady gas temperature at the combustor exit is acquired at 100 samples per second using a National Instruments based USB card via LabView software. The NO_x, CO and CO_2 emissions are measured using an electrochemical sensor based gas analyzer (KANE make) and NDIR sensor based gas analyzer (AIRSON make) for estimating the nitric oxide emission index at the combustor exit.

2.5 RESULTS AND DISCUSSION

2.5.1 FLAME REGIMES AND FLAME LENGTH

The visible appearance of swirl IJF stabilized in the quartz tube confined combustor is investigated to understand the effect of air jet aerodynamic swirl numbers on the flame shape, soot radiation luminosity and flame height qualitatively.

The visible flame appearance of quartz tube confined swirl IJF with the variation in the aerodynamic swirl number is shown in Figure 2.4 for a constant fuel mass flow rate of 7.1×10^{-2} g/s. The presence of an inner blue flame enveloped by an outer orange flame is evident for the SN_a of 9.54 as shown in Figure 2.4. The weak swirl generated by low tangential air mass flow rate at the coaxial burner exit results in poor entrainment and mixing of the annular fuel jet with the swirling central air jet. In addition, the absence of ambient air entrainment due to quartz tube confinement results in partial burning of the fuel jet in rich mode leading to the formation of a luminous outer flame. Interestingly, luminosity of the outer flame is found to diminish drastically along with the visible flame height with an increase in SN_a from 9.54 to 13.83 (see Figure 2.4).

The variation of visible flame height of swirl IJF with SN_a is reported in Figure 2.5 for two fuel mass flow rates. The normalized visible flame height of the confined swirl flame exhibits an exponential decay and it reduces by roughly 60% with an increase in SN_a for the fuel mass flow rates of $7.1 \ 10^{-2}$ g/s and 8.76×10^{-2} g/s respectively. The increase in tangential air mass flow rate for a fixed fuel mass flow rate in this combustor configuration establishes an inner recirculation zone (IRZ) close to the burner exit which helps in the effective entrainment and mixing of fuel jet with the swirling air jet. This aspect helps in establishing a nonluminous swirl IJF with compact reaction zone as evident from both the visible appearance and visible flame height.

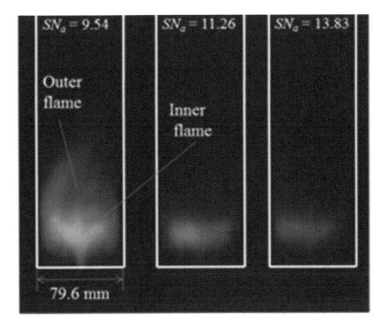

FIGURE 2.4   Instantaneous snapshots of the confined swirl IJF at different air jet aerodynamic swirl numbers for fuel mass flow rate $(\dot{m}_f) = 7.1 \times 10^{-2}$ g/s.

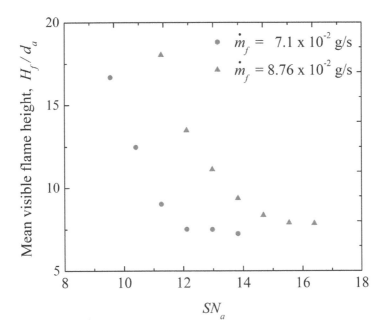

FIGURE 2.5   Variation of the normalized visible flame height of confined swirl IJF with aerodynamic swirl numbers for two fuel mass flow rates.

2.5.2   FLAME STABILITY AND BLOWOUT CHARACTERISTICS

The flame stability limits of the swirl IJF within the combustor are obtained by decreasing the fuel mass flow rate gradually for a fixed air jet aerodynamic swirl number. This study helps in identifying a demarcating boundary where the flame can sustain itself in the swirl flow field. The line in Figure 2.6(a) corresponds to the minimum fuel flow rate that is required for stabilizing the flame in this combustor for a fixed aerodynamic swirl number. The flame cannot be stabilized below this minimum fuel flow rate for a particular aerodynamic swirl number. It can also be observed that the minimum fuel mass flow rate required for stabilizing the flame in this combustor increases with an increase in the air jet aerodynamic swirl number.

The dynamics associated with the confined swirl flame in the proximity of lean blowout is discussed here. The reduction in the overall equivalence ratio (Φ) towards lean limit caused by a reduction in the fuel flow rate for a fixed air jet aerodynamic swirl number results in the decline of local heat release rate and hence the local flame speed. As a consequence, the flame cannot withstand the strain rate induced by the swirl flow field within the combustor. Subsequently, local extinction of the swirl IJF within the combustor takes place momentarily which can be observed from Figure 2.6b(iv). Due to this momentary local extinction, the fresh unburned fuel-air mixture pockets gets accumulated within the combustor. Depending on the

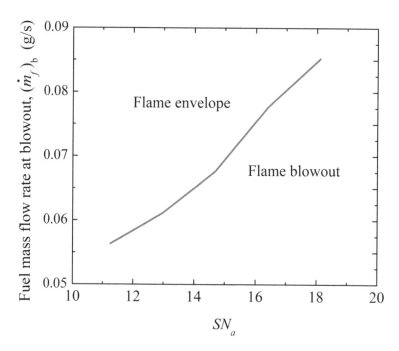

FIGURE 2.6(A)   Lean blowout limits of swirl IJF with aerodynamic swirl numbers in the labscale combustor.

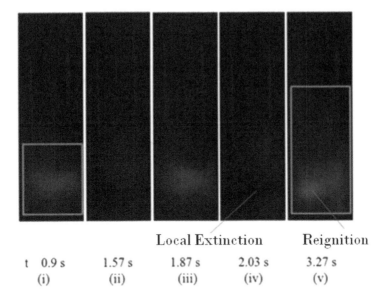

Local Extinction   Reignition

t   0.9 s   1.57 s   1.87 s   2.03 s   3.27 s
  (i)   (ii)   (iii)   (iv)   (v)

FIGURE 2.6(B)   Local extinction and reignition events in the swirl IJF based combustor for $SN_a = 18.11$ and $\dot{m}_f = 8.76 \times 10^{-2}$ g/s.

local fuel-air ratio of the accumulated fuel-air pockets, the hot product gases within the inner recirculation zone may ignite them intermittently which then leads to the probable reignition of a locally extinct flame region (see Figure 2.6b(v)). However, when the fuel flow rate is reduced beyond a threshold value for a fixed aerodynamic swirl number, the flame cannot generate enough heat release to reignite the accumulated air-fuel pockets, eventually leading to its global blowout within the combustor.

2.5.3   THERMAL AND NO_x EMISSION CHARACTERISTICS

The radial temperature and the post flame emission measurements are performed at the combustor plane which is 5 mm below the quartz tube exit to avoid ambient entrainment. The exit temperature profiles for two aerodynamic swirl number cases ($SN_a = 12.97$ and 16.4) for a fixed fuel mass flow rate of 8.76×10^{-2} g/s is shown in Figure 2.7.

A relatively flat temperature profile with a magnitude of around 1000 K is realizable at the combustor exit plane for the two aerodynamic swirl number cases. A dip in the temperature values close to the quartz tube walls can be attributed to the prevalence of heat losses from the quartz tube to the ambient air.

The post combustion NO_x emissions are measured in this swirl IJF based combustor by maintaining a constant fuel mass flow rate of 8.76×10^{-2} g/s and varying the aerodynamic swirl number gradually. This results in the variation of an overall equivalence ratio from 0.97 to 0.62 at the coaxial burner exit. The variation of an emission index of nitric oxide ($EINO_x$) with overall equivalence ratio is shown in Figure 2.8. The emission index (EI) of NO_x exhibits a monotonic decrease from

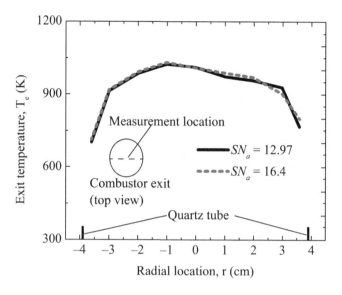

FIGURE 2.7   Radial temperature profiles at the swirl IJF combustor exit for two aerodynamic swirl number cases with $\dot{m}_f = 8.76 \times 10^{-2}$ g/s.

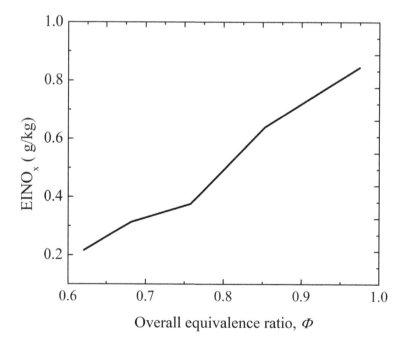

FIGURE 2.8   Variation of NO_x emission index with overall equivalence ratio in the swirl IJF based combustor.

0.84 g/kg to 0.21 g/kg with the reduction in overall equivalence ratio towards lean limit. With an increase in the aerodynamic swirl number, the air jet swirl velocity gets augmented which results in the formation of toroidal shaped inner recirculation zone (IRZ) close to the burner exit. The establishment of IRZ helps in the entrainment and the partial premixing of fuel jet with the central air jet. In addition, the presence of the small volume formed due to the recessing of the central tube in the coaxial burner as seen from Figure 2.3 helps in the dynamic partial premixing of swirl air jet and the annular fuel jet. An increase in the air jet aerodynamic swirl number for a fixed annular fuel flow rate is expected to enhance the extent of this dynamic partial premixing thereby supporting the reduction of the NO_x emission index in this combustor. In addition, the reduction in the flame temperature brought about by the lean overall equivalence ratio also favours the reduction of $EINO_x$ values in this IJF based swirl combustor. Earlier, Chen et al. [13] reported that an enhancement in the extent of fuel-air partial premixing favours reduction of $EINO_x$ values in the case of partial premixed methane swirling flame configuration.

2.6   CHAPTER SUMMARY

This chapter presented an overview of a typical gas turbine combustor and discussed the advantages of employing lean combustion in combustors from the perspective of pollutant emissions. Furthermore, it discussed the safety issues encountered in employing lean combustion in gas turbine combustors. Importantly, it introduced a distinct new nonpremixed flame configuration called an inverse jet flame which has the capability to establish a lean nonpremixed flame devoid of flame flashback and lean blowout. The experimental characterization of this swirl IJF based combustor in terms of visible appearance, flame height, flame stability, exit temperature distribution and nitric oxide emission levels at the exhaust clearly reveals that it is possible to establish and anchor a compact blue flame with superior thermal and emission characteristics without flame flashback. The preliminary results discussed in this chapter reveal that the inverse jet flame based swirl combustor can be considered as a potential candidate for adaption in gas turbine engines.

REFERENCES

[1] Lefebvre, A.H., & Ballal, D.R. (2010). *Gas Turbine Combustion—Alternative Fuels and Emissions*, Third Edition, CRC Press, Boca Raton, FL.

[2] Mishra, D.P. (2010). *Fundamentals of Combustion*, Revised Edition, Prentice Hall of India, New Delhi.

[3] McAllister, S., Carlos Fernandez-Pello, A., & Chen, J.Y. (2011). *Fundamentals of Combustion Processes*, Springer, New York.

[4] Lee, T.W., Fenton, M., & Shankland, R. (1997). Effects of partial premixing on turbulent jet flame structure. *Combustion and Flame*, *109*, 536–548.

[5] Sobiesiak, A., & Wenzell, J.C. (2005). Characteristics and structure of inverse flame of natural gas. *Proceedings of Combustion Institute*, *30*, 743–749.

[6] Sze, L.K., Cheung, C.S., & Leung, C.W. (2006). Appearance, temperature and NO_x emission of two inverse diffusion flames with different port design. *Combustion and Flame*, *144*, 237–248.

[7] Mahesh, S., & Mishra, D.P. (2008). Stability and emission characteristics of turbulent LPG IDF in a backstep burner. *Fuel, 87,* 2614–2619.

[8] Mahesh, S., & Mishra, D.P. (2019). Effect of air jet momentum on the topological features of turbulent CNG inverse jet flame. *Fuel, 153,* 1068–1075.

[9] Mahesh, S., & Mishra, D.P. (2015). Flame stability limits and near blowout characteristics of CNG inverse jet flame. *Fuel, 153,* 267–275.

[10] Mahesh, S., & Mishra, D.P. (2015). Dynamic sensing of blowout in turbulent CNG inverse jet flame. *Combustion and Flame, 162,* 3046–3052.

[11] Mahesh, S., & Mishra, D.P. (2015). Characterization of swirling CNG inverse jet flame in recessed coaxial burner. *Fuel, 161,* 182–192.

[12] Sautet, J.C., Boushaki, T., & Labegorre, B. (2009). Control of flames by tangential jet actuators in oxy-fuel burners. *Combustion and Flame, 156,* 2043–2055.

[13] Chen, R.H., & Driscoll, J.F. (1998). The role of the recirculation vortex in improving fuel-air mixing within swirling flames. *Proceedings of Combustion Institute, 22,* 531–540.

3 Burner Technology for Hydrogen Fuel

Debi Prasad Mishra and Swarup Y. Jejurkar

CONTENTS

3.1 INTRODUCTION

3.1.1 PROPERTIES OF HYDROGEN

Hydrogen (H_2) is an energetic material used as a source of thermal energy due to its advantageous thermochemical properties [1]. Hydrogen is a highly diffusive gas under the action of concentration gradients. It is also highly diffusive under the action of thermal gradients, with a tendency to accumulate faster in the regions of high temperature. Combustion of hydrogen produces water vapor and releases 120 MJ/kg (lower heating value) in the process. The released heat is significantly more than many hydrocarbons (~45 MJ/kg).

Hydrogen also has wider flammability limits than most of the fuels, with lower and upper flammability limits of 4 and 75 vol % in air, respectively. Thus, very weak flames of hydrogen are also observed. The quenching diameter (d_q) of hydrogen is also small (d_q ~0.6 mm), so that very weak hydrogen flames could be supported in small cracks [2]. The burning intensity of hydrogen is also high and laminar burning velocity of stoichiometric H_2-air flame is ~2.8 m/s and ~14 m/s when O_2 is used.

Thermophysical and thermochemical properties of hydrogen mentioned previously make it one of the most important fuels in combustion technology. Hydrogen is well known for its application as a cryogenic liquid propellant where it is paired

DOI: 10.1201/9781003049005-3

with oxygen. Hydrogen has many more applications in industrial combustion and gas turbines. We consider some applications for hydrogen combustion in these areas.

3.1.2 ORGANIZATION OF CHAPTER

In this chapter, we consider the combustion of hydrogen with emphasis on its storage methods and burner technology.

In Section 3.2, an overview of hydrogen storage technologies is presented. Limitations encountered in hydrogen storage are discussed first and storage methods potentially useful in hydrogen combustion applications are then considered. Some of the new developments are also briefly surveyed. In Section 3.3, hydrogen burner systems are considered with a description of three burner systems. Use of hydrogen in its pure state is considered. The section includes design principles, a description of burners, and their operational characteristics. A brief perspective of the historical developments is also presented in this section. We conclude with some remarks on the scope for the future in Section 3.4.

3.2 HYDROGEN STORAGE TECHNOLOGIES

A chief impediment in the application of hydrogen for combustion technologies is the difficulty encountered during its storage. Significant progress in storage technologies could facilitate a widespread application of hydrogen as fuel in combustion systems. Therefore, we discuss some features of the storage technologies considered for hydrogen in this section.

3.2.1 LIMITATIONS IN HYDROGEN STORAGE

Hydrogen (H_2) is the lightest diatomic element, with a molecular weight of 2.016. Apart from the covalently bonded diatomic form, atom (H), anion (H^-), and cation (H^+) states are also known to occur in compounds. These compounds are "hydrogen carriers."

Figure 3.1a is the phase diagram for H_2 in the pressure (P) and temperature (T) space. It shows that H_2 is in a solid state at very low temperatures, with a density (ρ) of 70.6 kg/m³ at 10 K [3]. It passes through a semi-solid "slush" state as temperatures rise into a liquid state under restricted P and T range. Liquid H_2 has a density of 70.8 kg/m³ at 20 K [3]. H_2 is a gas under normal temperature and pressure (NTP) conditions with a significantly decreased density of ~0.089 kg/m³ at 273 K [3]. A metallic state is reached at very high pressures and/or temperatures as indicated in Figure 3.1a. As far as storage is considered, density, pressure, and temperature are critically important. Unfortunately, storage pressure and temperature are extreme and very difficult to maintain for a considerable time.

Moreover, hydrogen is not dense enough (1 kg H_2 needs 11 m³ at NTP [3]), which greatly restricts the amount of hydrogen that could be stored in a given volume. Density variations are shown in Figure 3.1b. An important consideration here is the requirement of thick-walled cylinders as compression pressures increase. This requirement is expressed in Figure 3.1b in terms of the ratio of wall thickness (t_w) and

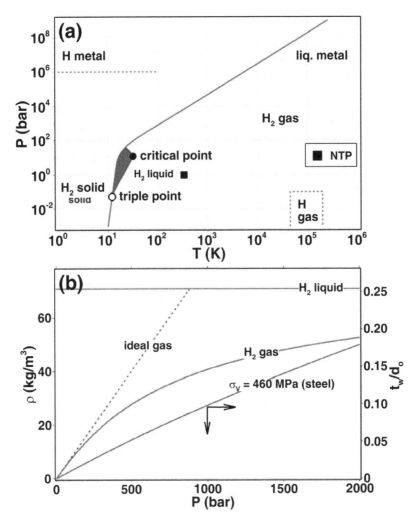

FIGURE 3.1 (a) Phase diagram for hydrogen [3] and (b) density variations and required geometric properties of a steel cylinder storing hydrogen [3].

outer diameter (d_o). Hydrogen storage in cryogenic conditions is a method adopted for rocket propulsion applications. The liquid hydrogen state is deemed essential for achieving the required performance of an engine and to achieve adequate storage volumes for the duration of mission. However, it is very difficult to maintain hydrogen in liquid state in large volumes.

Therefore, the possible avenues for storing H_2 in its free state are limited to compression or liquefaction. In the chemically combined state, many more possibilities exist. However, covalently bonded chemical compounds (typically involving C-H bonds) are less desirable because hydrogen release from such compounds requires decomposition or combustion level temperatures (~800–1000 K) and severely limits their energy economy.

>50**   Advances in Combustion Technology

3.2.2 Storage Methods

Figure 3.2 summarizes the many methods for storing H_2 in a free or combined state. Also given are the maximum uptake or gravimetric density (in mass %), density (in kg/m³), and typical storage temperature (in K) and pressure (in MPa). All the listed methods are potentially useful when hydrogen is used as a fuel in combustion.

High-pressure compression (~20 MPa) is the most common method of storing free hydrogen. Innovations in pressure vessel technology and steelmaking enable storage pressures as high as 80 MPa [3] and densities ~40 kg/m³, as shown in Figure 3.1b. At the same time, required strength of cylinder walls goes up and it diminishes the gravimetric density [3]. Composite materials of much higher tensile strength (σ_y) than steel and less density are being considered for high-pressure compression of H_2 gas [3]. Innovations are also underway to improve the safety margin for large-scale hydrogen storage using compression [3]. However, inherently low achievable density and relatively very high pressures are the important bottlenecks for compressed hydrogen storage systems.

Even the best technologies for compressed H_2-gas storage only offer density approximately half (~30 kg/m³) of the liquid H_2 (~70 kg/m³), chiefly due to limitations on the structural integrity of storage vessels. With reference to Figure 3.1a, liquid H_2 is essentially a cryogenic fluid at 21.2 K and ambient pressure. It is stored in an open system due to the low critical point of hydrogen (33 K). Massive over-pressures

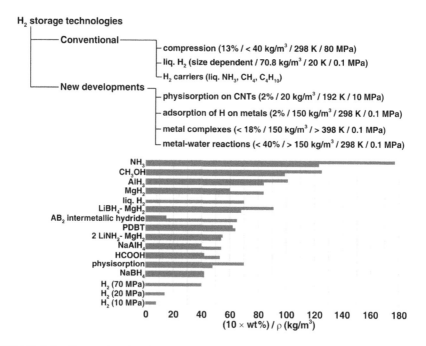

FIGURE 3.2 Conventional technologies for storing hydrogen and new developments [3] along with current capability of technologies to store hydrogen in terms of stored mass and mass density [4].

could be generated in a closed system [3]. Developments in liquefaction and insulation for cryogenic tanks have resulted in highly sophisticated storage systems that are routinely used for giant liquid rocket engines via on-board storage. The liquefaction process is complicated by the presence of para- and ortho-hydrogen and proper control of the process ensures that vaporization would be reduced [3]. The Linde (or Joule-Thompson) cycle is used for liquefaction as a first approximation.

Vaporization (also called boil-off) depends on the size and shape of the tank and insulation used [3]. The large size and spherical shape of the pressure vessel would reduce the boil-off. Cryogenic storage systems are expensive and suffer from inevitable losses in stored mass. Therefore, their application in combustion is exclusively for rocket propulsion. Nevertheless, cryogenic storage systems could also be used for hydrogen carriers, chemical compounds containing a large fraction of hydrogen atoms bonded to N and C atoms. Thus, we could use liquid ammonia (NH_3), methane (CH_4), and higher hydrocarbons (e.g. butane, C_4H_{10}). Sometimes, H_2-carriers are also called chemical hydrides [4]. Liquid NH_3 contains 17.7% hydrogen by gravimetric and 123 kg/m³ by volumetric proportions at 1 MPa [4].

Figures 3.2 and 3.3 indicate the achievable volumetric and gravimetric densities for compression- and liquefaction-based conventional storage systems for hydrogen

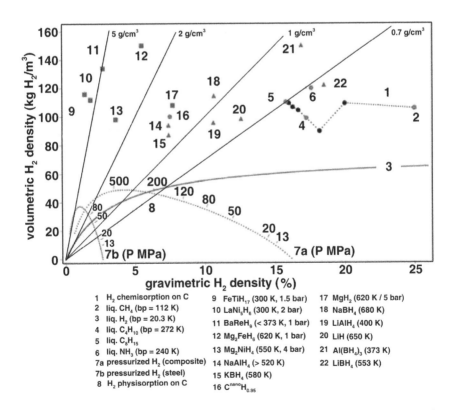

FIGURE 3.3 Correlation between volumetric and gravimetric densities achievable using many technologies for storing hydrogen in gas, liquid, or solid states [3].

fuel. The diagrams also show the performance of some of the new developments taking shape in this area.

3.2.3 NEW DEVELOPMENTS

New technologies for hydrogen storage are based on adsorption and hydrogen carriers. Adsorption-based technologies include the ones using physisorption of H_2 on carbon nanostructures (tubes and sheets). In case of hydrogen, weak interactions limit physisorption to sub-zero temperatures [3]. On a single-sided graphene sheet of specific surface area ~1315 m²/g, maximum 3% adsorption in monolayer is possible [3]. Tubes and microporous capillaries offer benefits of overlapping attraction forces due to the high curvature [3]. Figure 3.4 shows the maximum uptake in mass % for carbon nano tubes as a function of the specific surface area (a). Figure 3.4 also shows that the adsorbed mass of hydrogen is proportional to the specific surface area at 77 K.

Apart from carbon, zeolites, glass microspheres, and capillaries are also being explored [3, 5]. Low operating pressure, cheaper materials, and simple designs are the important advantages of the physisorption-based storage systems when compared against the relatively low uptake and low temperatures [3].

Among the new developments, the most prominent presence of metal hydrides can be noticed in Figures 3.2 and 3.3. Hydrides are compounds of hydrogen with metals, intermetallic complexes, and alloys. Figures 3.2 and 3.3 also show that the highest volumetric densities are achieved by metal hydrides; for example, $\rho = 150$ kg/m³ for Mg_2FeH_6 and $Al(BH_4)_3$ [3, 4].

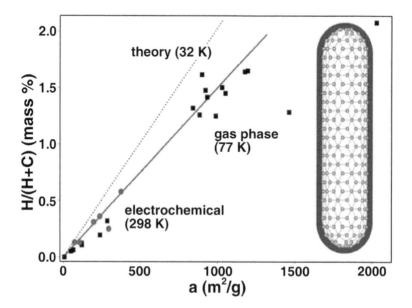

FIGURE 3.4   Physisorption of hydrogen on carbon nano tubes [3].

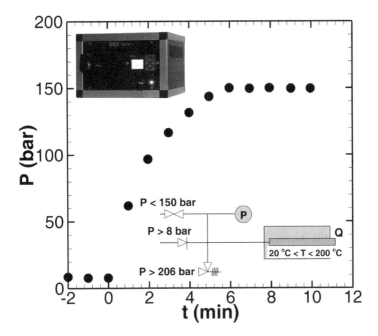

FIGURE 3.5 Metal hydride compressor, its control mechanism, and operating characteristics [6].

Among the metal hydrides, compounds with elemental metals include magnesium hydride (MgH_2) and aluminum hydride (AlH_3). MgH_2 requires high temperatures above 573 K for activation to overcome the thermodynamic and kinetic restrictions [4]. While storage could be as high as 7%, this is still a developing field. AlH_3 bonds hydrogen rather weakly and could enable densities as high as 10% at 373 K. However, manufacture (also called regeneration) of AlH_3 is very difficult [4]. Intermetallic hydrides and their processing are expensive [4]. Alanates are complex metal hydrides in which hydrogen is part of a complex (AlH_4^-) anion bonded with a cation (e.g. Na^+). Due to the presence of light elements in alanates, gravimetric hydrogen capacities increase, although high temperatures are required for their activation [4].

Alternative to the conventional compressed gas cylinder, a metal hydride compressor (HyCo) has been developed [6]. It is shown in the inset of Figure 3.5. It adsorbs hydrogen from a low-pressure source at $P < 0.8$ MPa. The bed is then heated up to the required desorption pressure. It could reach 15 MPa in less than 6 min, as shown in Figure 3.5.

The control system is also shown in Figure 3.5. Check valves ensure the correct flow direction and isothermal desorption is continued. Adsorption begins once the cooling and depressurization occurs. HyCo is capable of supplying gaseous H_2 at 400 nml/min and 15 MPa. It could store 70 nl hydrogen in a package of $350 \times 250 \times 170$ mm^3 with a starting pressure of 0.8 MPa [6]. Development of the metal hydride compressor has progressed towards higher pressures towards 50 MPa [6] using different metal hydrides.

The metal hydride–based storage systems are evolving to become as simple as possible while obeying the constraints of weight, volume, pressure, and temperature so that required flow rates are obtained [6]. Safety and reliability aspects of these systems have also received attention [5].

3.3   HYDROGEN BURNER SYSTEMS

3.3.1   HISTORICAL PERSPECTIVE

Use of hydrogen in a pure state for combustion has been claimed for a very long time [7]. Modern efforts explored the use of hydrogen in aircraft gas turbines [8]. JP-5 and JP-6 mixed with up to 15% H_2 were shown to combust with widened lean blowout limits. Concomitant drop in the emission of NO_x to less than 10 ppm at 15% O_2 and reasonably low levels of CO and unburnt hydrocarbons were also achieved in these tests. Demonstrations of the effects of hydrogen also included premixing with propane [9]. These investigations extended the lean flammability limits and improved the combustion efficiency at inlet temperatures and pressures as high as 700 K and 5 MPa. Remarkably diminished emission of NO_x (0.06 ppm at 98% efficiency and $\phi = 0.24$) was also demonstrated using premixed hydrogen and air [10].

3.3.2   DESIGN PRINCIPLES AND BURNER SYSTEMS

Hydrogen combustion in a gas turbine has been investigated in two burner configurations [11]. The concepts of these burners are illustrated in Figure 3.6. In one configuration, non-premixed flame is considered while another considered a multi-injector operation.

The injection pattern was of the coaxial type with an inverse contacting pattern of oxygen inner jet surrounded by outer hydrogen jet. An inverse pattern prevents hot O_2 from affecting the liner wall material. Both configurations used slightly fuel-rich ($\phi > 1$) or stoichiometric ($\phi = 1$) combustion and a mixing region. A liner was used to separate combustion and mixing regions [11]. These burners are characterized as steam generators. The choice of hydrogen is obvious once the absence of SO_x, CO_x, and NO_x is considered. They have been designed to combust a stoichiometric H_2-O_2 mixture in steam at 773 K and produce a high-pressure steam at 1973 K and 5 MPa.

In Figure 3.6a, a pilot burner is shown to combust at slightly fuel rich condition along with multiple injectors for H_2. A pilot flame is used to stabilize the non-premixed flame. In Figure 3.6b, the multi-coaxial injector is used to obtain stoichiometric premixed flame. The injectors contain recessed O_2 to effectively mix two gases of widely different molecular weights [11].

Schefer et al. [12] evaluated lean premixed hydrogen combustion, with potential application in aircraft gas turbines for ultra-lean combustion and elimination of unburnt hydrocarbons and CO_2. As shown in Figure 3.7, nine channels were used to supply air (maximum Mach number of 0.7). Hydrogen was injected radially inward

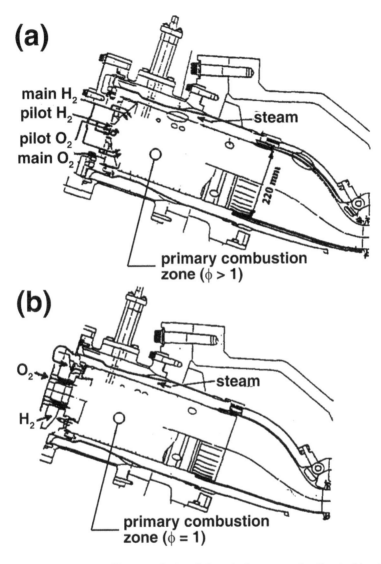

FIGURE 3.6 Two types of burners designed for a hydrogen combustion turbine [11]; (a) non-premixed flame and (b) multiple coaxial injector configurations.

(i.e. in crossflow) through two orifices located on either side of each channel (inset of Figure 3.7). These orifices were fed via four tubes at the injector base plate and circulate inside the base plate to achieve cooling. Mixing was achieved in a compact mixing zone. Uniform mixing could be achieved for a high momentum ratio between fuel and primary air. Lean combustion was considered to reduce the NO_x emission and still obtain a stable flame using high burning velocity of hydrogen and resistance to stretch.

FIGURE 3.7 NASA premixed flame burner using hydrogen [12] and details of hydrogen and air injection ports.

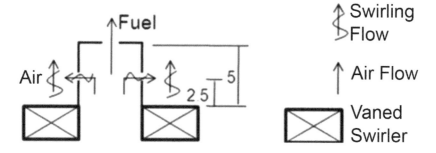

FIGURE 3.8 Crossflow configuration for hydrogen-air combustion using swirling combustion air [13]. All dimensions are in mm.

Mishra and Jejurkar [13] used a crossflow configuration, as shown schematically in Figure 3.8. Hydrogen fuel is injected externally into the combustion air through three orifices arranged circumferentially on the lateral surface of the burner head. One centric orifice is also used in addition to the circumferential ones. The radial outflow of fuel jets is directly into the path of swirling combustion air arranged around the burner head. Radial fuel jets exit 2.5 mm downstream and the centric fuel jet exits 5 mm downstream of air exit, respectively. It is expected that such an arrangement would result in fast mixing of fuel and air external to the burner head and the flame stability envelope might be widened.

3.3.3 OPERATIONAL CHARACTERISTICS

The concept of burners intended for a steam turbine based on hydrogen combustion was realized at the laboratory scale in the instrumented burner [11], as shown in Figure 3.9. It used one coaxial injector. O_2 was supplied from the first port on the left of Figure 3.9a, H_2 from the middle port, and steam from the right port. A converging nozzle was used to maintain the chamber pressure and prevent reverse

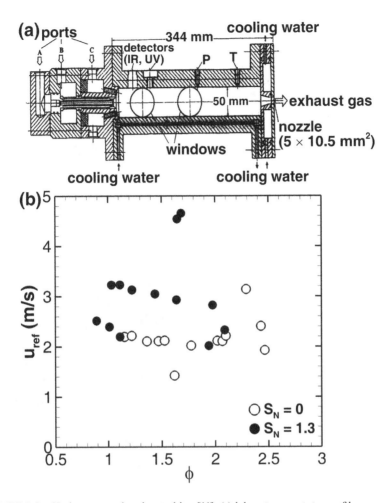

FIGURE 3.9 Hydrogen combustion turbine [11]; (a) laboratory prototype of burner and (b) operating map for burner characterization without steam.

flow from ambient. An O_2 jet is swirled with geometric swirl number (S_N) of 1.3 to improve mixing.

Injection velocity for the jets is 100–250 m/s to prevent flashback of the premixed flame of a high burning velocity (max. of 14 m/s). Figure 3.9b shows the range of conditions for which stable flames were obtained. Reference velocity (u_{ref}) is used to denote flow conditions averaged over the cross-section of the combustion chamber. Equivalence ratio (ϕ) range of 0.9–2.0 was obtained for the stable flames, irrespective of the swirl. However, swirl extended the operation to lower u_{ref} and fuel-rich ϕ. Additionally, the shortening of flame length is another factor in favor of the swirling O_2 jet. Hence, swirl was considered better.

Experiments on H_2-O_2 steam were conducted using wet steam at 423 K and 1 MPa and indicated stable combustion as shown in Figure 3.10, probably due to the

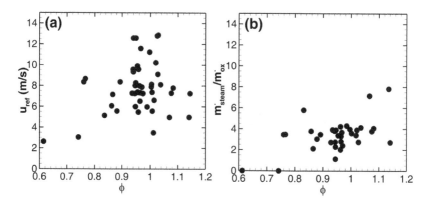

FIGURE 3.10 For hydrogen combustion turbine [11]; (a) stable operating conditions in velocity-equivalence ratio coordinates when using steam and (b) stable operating conditions in terms of steam-oxygen mass ratio and equivalence ratio.

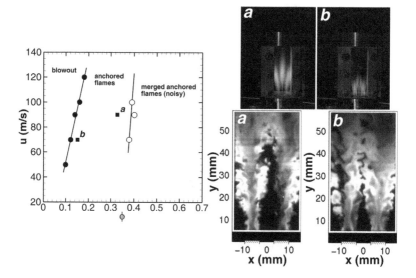

FIGURE 3.11 Stability characteristics of the premixed hydrogen flame burner [12]. Inset shows the visible and OH-PLIF images for two examples of the regime of multiple anchored flames.

separate mixing and combustion regions. However, the correlations between u_{ref} and ϕ are not observed in Figure 3.10a. The dilution ratio in terms of ratio of mass flow rates of steam and O_2 shows that stable combustion was achieved for significant dilution levels also, as shown in Figure 3.10b. Thus, same burner designs could be useful for H_2-O_2, as well as highly diluted H_2-O_2-steam combustion.

In Figure 3.11, the stability characteristics of the premixed flame of hydrogen and air are shown in terms of bulk inlet velocity (u) and overall equivalence ratio (ϕ). The burner used for these experiments is the same as shown in Figure 3.7.

Three flame regimes were observed; merged anchored flame, multiple anchored flames, and the unstable blowout. All flames were anchored to the air channels. Merged anchored flame occurred at high equivalence ratios and low inlet velocities. They were noisy and could be flashback-stabilized flames, probably due to the competition between higher flame speed and lower inlet velocity [12]. The noise is characterized as a high-pitched and loud whistle. From the safety point of view, these conditions are to be avoided and hence the middle anchored flame regime is the most desirable. Unstable lifted flames have been speculated for $u > 120$ m/s [12]. Operating conditions denoted by a and b are located in the second regime and visualized using natural luminosity as well as OH-PLIF in the adjacent images. Flames near blowout are short and less luminous. The OH-PLIF scans show the zone of high OH concentration extending to the inlet region [12] along the interface between the jet and recirculation regions. Moreover, higher momentum of fuel is favorable for mixing as indicated by acetone-PLIF [12].

NO$_x$ emissions from the crossflow burner configuration shown in Figure 3.8 were compared against the single orifice design in which only the centric orifice ($D = 4$ mm) was retained and lateral orifices responsible for the crossflow were absent. Emission levels reduced with the use of crossflow design in the range of ϕ_{global} of interest, as shown in Figure 3.12. The arrangement of lateral orifices in addition to the centric orifice of some conventional configurations significantly reduced NO$_x$ emissions in hydrogen combustion. This is achieved by improved mixing in the

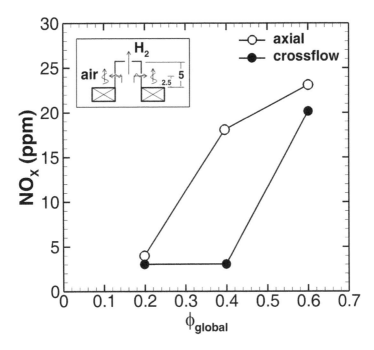

FIGURE 3.12 Performance of the crossflow and axial flow schemes of hydrogen injection in terms of NO$_x$ emission [13].

crossflow arrangement and elimination of hot gas pockets in the flame zone, along with lean combustion.

The full factorial design method was instrumental in establishing an empirical correlation of NO_x emission (y, in ppm) with the global equivalence ratio (ϕ_{global}) and desired power level (p)—viz., $y = 6.03125 - 16.25\varphi_{global} - 3.01667p + 15.25\varphi_{global}p$. Only a small number of experimental runs were needed to arrive at the relationship and operating parameters were varied simultaneously, unlike the traditional method in which only one operating parameter is varied at a time. The results of experimental program are summarized in Figure 3.13 [13].

This correlation could be useful in evaluating the performance of burner in a future design activity. Global equivalence ratio is the dominant factor in determining

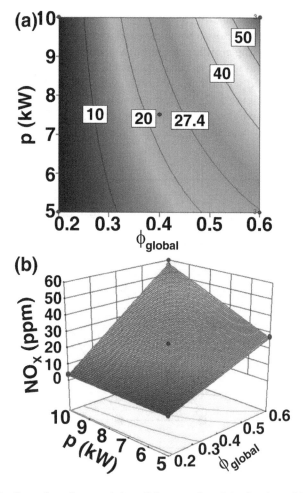

FIGURE 3.13 Operating characteristics of the crossflow injection in swirling stream of combustion air in terms of NO_x emission (in ppm) with operating conditions expressed as H_2-equivalent power input and global equivalence ratio [13].

emission level, although the combined effects of power level and global equivalence ratio are on the same order as the power level. A useful operating region for this kind of burner is in the range of low enough global equivalence ratios, irrespective of the power level since the emissions are reduced in this range.

3.3.4   CHALLENGES

Storage technologies for hydrogen fuel have reached a stage where further development calls for radical shifts. While intensive efforts being made in this area have resulted in the emergence of many competing methods that seek to increase the uptake of hydrogen at relatively mild pressure and temperature, they are yet to mature to the level of conventional methods. In the short term, therefore, the emphasis in hydrogen burner technology would be towards the use of compressed gas cylinders. Urgent efforts are now required in the integration of storage and burner technologies with emphasis on improvement in safety and reliability of conventional compressed gas storage systems.

Some of the described burner technologies are inspired by the use of hydrogen in rocket engines. However, due to cost considerations, significant compromises must be made to make the burner technology affordable and reliable. Nevertheless, high temperature reached in the combustion chamber demands the development of thermal insulation around the primary combustion region [11]. Other difficulties have been listed, including a flow control mechanism for maintaining stoichiometric proportions, a cooling mechanism, and reliable flame detection in high-temperature steam [11] so that combustion of hydrogen could be successfully integrated with a turbine.

3.4   CONCLUDING REMARKS

Burner technology relying on hydrogen as a fuel is under continuous development. However, paltry progress has been made in the experimental demonstration of concepts, stabilization of flames, and characterization of burner properties. Hence, in this chapter, burner technology for hydrogen fuel is presented with a focus on flow design principles, operational characteristics, and flame appearance and structure for a few burner systems. The achievements so far suggest that burners can exploit effectively the favorable thermochemical properties of hydrogen to achieve reduced NO_x emission, wide stability limits, and sustained combustion of hydrogen. However, successful deployment of these burners in the applications is still plagued by deficiencies in the storage systems. Except for the metal hydride compressors and cryogenic storage systems, other storage technologies require more development for use alongside a hydrogen combustion system.

REFERENCES

1. I. Glassman and R.A. Yetter. *Combustion.* 4th edition, Academic Press, Cambridge, MA, 2008.
2. V.R. Lecoustre, P.B. Sunderland, B.H. Chao, and R.L. Axelbaum. Extremely weak hydrogen flames. *Combust. Flame* 157 (2010), pp. 2209–2210.

3. A. Züttel. Hydrogen storage methods. *Naturwissenschaften*. 91 (2004), pp. 157–172.

4. J. Andersson and S. Grönkvist. Large-scale storage of hydrogen. *Int. J. Hydrogen Energy* 44 (2019), pp. 11901–11919.

5. R. Moradi and K.M. Groth. Hydrogen storage and delivery: review of the state of the art technologies and risk and reliability analysis. *Int. J. Hydrogen Energy* 44 (2019), pp. 12254–12269.

6. J.B. von Colbe, J.-R. Ares, J. Barale, et al. Application of hydrides in hydrogen storage and compression: Achievement, outlook and perspectives. *Int. J. Hydrogen Energy* 44 (2019), pp. 7780–7808.

7. R.A. Erren. Internal Combustion engine using hydrogen as fuel. US 2,183,674A, 19 December 1936.

8. R.M. Clayton. Reduction of gaseous pollutant emissions from gas turbine combustors using hydrogen-enriched jet fuel—a progress report. NASA TM 22–790, 1976.

9. D.N. Anderson. Effect of hydrogen injection on stability and emissions of an experimental premixed prevaporized propane burner. NASA TM X-3301, 1975.

10. D.N. Anderson. Emissions of oxides of nitrogen from an experimental premixed-hydrogen burner. NASA TM X-3393, 1976.

11. T. Hashimoto, K. Koyama, and M. Yamagishi. Hydrogen combustion characteristics in a model burner with a coaxial injector. *Int. J. Hydrogen Energy* 23 (8) (1998), pp. 713–720.

12. R.W. Schefer, T.D. Smith, and C.J. Marek. Evaluation of NASA lean premixed hydrogen burner. SAND2002–8609.

13. S.Y. Jejurkar and D.P. Mishra. Characterization of confined hydrogen-air jet flame in a crossflow configuration using design of experiments. *Int. J. Hydrogen Energy* 38 (2013), pp. 5165–5175.

4 Flameless Combustion with Liquid Fuels for Ultra-Low Emissions from Combustion Systems

Saurabh Sharma and Sudarshan Kumar

CONTENTS

4.1 INTRODUCTION

Gas turbine combustors have been of extreme importance in air transportation and power generation. Their use has been increased over the last two decades and will be the prime source for air transportation for the next couple of decades, keeping in mind the less efficient power storage systems such as batteries. It is, therefore, crucial to reduce the harmful gaseous and noise levels from the existing gas turbines. It is proposed in the Flightpath 2050, Europe's Vision for Aviation, that NO_x emissions are expected to be reduced by 90% and the noise of the flying aircraft by 65% compared to the typical aircraft in 2000 [1].

Demand has been of high thermal efficiency from the gas turbines; for that, the turbine inlet temperature and pressure ratio have been increased. However, it comes with its challenge of increased NO_x emissions due to high temperature. It is therefore required to maintain the low NO_x emissions along with high thermal efficiencies. Feasible solutions to achieve that feat are (i) use of renewable/alternative fuels,

(ii) change in the system design of the combustors, (iii) use of a new combustion technique. The latter seems to be a suitable choice due to its inherent potential.

MILD (moderate or intense low-oxygen dilution)/flameless combustion is an excellent choice to work with due to its exceptional qualities like well-distributed combustion, low gaseous emissions, and small acoustic oscillations. Flameless combustion is a non-stabilised, mixing controlled phenomenon in which the decoupling between the flow dynamics and the heat transfer leads to unique combustion characteristics. The most important feature among all is the low NO_x formation. It is the result of many important mechanisms in flameless combustion such as (i) a distributed reaction zone, (ii) reduced availability of the primary reactants (fuel and oxidiser), (iii) formation of NO_x via a prompt route rather than a thermal route. The low oxidiser concentration environment in the flameless combustion is achieved by the recirculation of the hot combustion products. In gas turbine combustors, it can be achieved through geometric modifications to create internal recirculation. Also, the flameless combustion has the advantage of a broad stability range which is a crucial factor for the gas turbines as they have to operate at part load very often. It is applicable for both land-based and aviation gas turbines. The stability at part load can be achieved using recirculation of hot combustion products. Also, the application of different alternative fuels makes the use of flameless combustion even more attractive due to the increased stability range.

Flameless combustion in industrial furnace and burners has been studied in the past several years, but its precise application in gas turbines has not been found yet. In recent years, several concepts and designs have been tested for the implementation of the flameless combustion to gas turbines. However, it has never been easy to completely mimic the operating conditions of the gas turbines due to following reasons:

1. In the existing design of can-type combustors, it is challenging to achieve the desired recirculation and dilution needed for flameless combustion.
2. The range of global equivalence ratio ϕ varies in the range of 0.3–0.4 in gas turbine combustors, which makes it challenging to achieve low oxidiser concentration to attain the flameless combustion mode.
3. Volume and weight requirements of the aircraft are very critical, which makes it challenging to achieve higher heat release densities in flameless combustion.

Kruse et al. [2] studied the MILD combustion in a reverse flow configuration for methane at different operating pressures. Reverse flow results in increased recirculation leading to low CO and NO_x emissions. Ye et al. [3] investigated the same test rig [2] with pre-vaporised liquid fuels at increased pressures. NO_x was found to increase with increasing pressure. However, the use of N_2 as carrier gas would lead to NO_x reduction [3]. Arghode et al. [4, 5] studied the colourless distributed combustion (CDC) in a reverse flow prismatic combustor at high intensities of 53–85 MW/m^3-atm for methane. The fuel and air were injected in cross-flow, and the emissions were minimised.

Khalil et al. [6, 7] investigated swirl stabilised colourless distributed combustors for the application in the gas turbine combustors at high heat intensity (2.4 MW/ m³-atm) with gaseous [6] as well as liquid fuels [7]. CO_2 and N_2 were mixed with the air to simulate the exhaust gases recirculation and reduced oxygen concentration. For pre-vaporised liquid fuels, CO_2 was added to fuel, and the reaction zone was studied through OH* chemiluminescence [7]. They concluded that the O_2 level at which the transition to colourless distributed combustion occurs is independent of the fuel type [7]. Sorrentino et al. [8] studied a cyclonic combustor with different inlet temperatures. The cyclonic motion was generated by injecting the air through opposite ends. Increased residence time by adopting the cyclonic flow field helped achieve complete combustion. Although, the high residence times (~0.5 sec) is not suitable for the typical gas turbine conditions [9].

Some preliminary work has been reported on the use of flameless combustion to gas turbines. Lückerath et al. [10] studied the hexagonal shaped FLOX® combustion at high pressure (~ 20 bar) with different fuel compositions. They found that high jet velocities lead to low NO_x and increased stability range [10]. Lammel [11] increased the thermal intensity of the FLOX® combustor. They adopted a novel mixing method called HiPerMix in which the reactants are mixed before entering into the combustion chamber. The NO_x emissions were reported low (~ 10 ppm) along with substantial pressure loss [11]. Sadanandan et al. [12] investigated the same FLOX® combustor with a separate arrangement for the different degrees of premixing at 20 bar pressure. They concluded that collectively, the degree of premixing and recirculation rate affects the NO_x and CO formation. They achieved very low emissions, but the operating range was very narrow. To extend the stability over a wide range of air-fuel ratio, Schütz et al. [13] developed the EZEE® configuration. They were able to maintain the stability in full load as well as part load by the radial staging of the fuel. These studies identified the high jet velocity as a critical factor to achieve low emission combustion [13].

In a first, Levy et al. [14] proposed the concept of a FLOXCOM combustor in which the desired recirculation is achieved using geometric modifications and air splitting. Melo et al. [15, 16] experimentally verified the findings of Levy et al. [14] by developing a prototype annular combustor. The fuel is injected in the already vitiated zone to ensure distributed reaction. They tried different air inlet configurations at different angles to optimise the recirculation. They concluded that recirculation is mainly dependent on geometry [16]. Measured NO_x was very low, but the level of CO and HC was too high for the gas turbine application. This issue was attempted by Levy et al. [17] through computational modelling. The fuel energy was released in two steps, and the air was also split. First, the fuel is injected in low O_2 region and heat was transferred to the secondary air stream. It helped in the reduction of both NO_x and CO.

Rao et al. [18] proposed a two-combustor approach to use flameless combustion in a gas turbine with liquid fuels. They used cryogenic fuel in the primary combustor and kerosene in the secondary combustor (flameless mode). It was proposed that the exhaust from the first combustor would have low O_2 concentration and will help to achieve the favourable conditions for flameless combustion in the secondary combustor. Recently, Levy et al. [19] studied a bubbly shaped secondary combustor for flameless combustion operation based on the concept of Rao et al. [18]. Air was divided into dilution and combustion streams. The combustion air was injected in the large recirculation zone to achieve low NO_x emissions.

Reddy et al. [20–23] studied the flameless combustion of liquid fuel in a swirl stabilised high heat intensities (21 MW/m³ ~85 kW) combustor. Tangential air injection was used to increase the residence time, and chamfered outlet was provided to enhance the recirculation inside the combustor. Fuel flexibility with the flameless combustion was investigated by Reddy et al. [23] and Sharma et al. [24]. Biodiesel showed different emission characteristic compared to kerosene and diesel [23]. Sharma et al. [25] studied the effect of the spray size on the combustion characteristics of the flameless combustor. They concluded that finer spray would lead to low gaseous and acoustic emissions. It is found from the literature that direct application of flameless combustion of liquid fuel in gas turbines lacks both technology and prototype. Although it is claimed to achieve in the secondary burner (burning kerosene) integrated with a conventional combustor (burning cryogenic fuel) [18], it is required to have a combustor operating in flameless mode with all technological advantages as a whole. The present study aims at the development of a flameless combustor operating in flameless mode with the novel geometric features. Initially, numerical simulations are carried out to obtain the flow field inside the combustor. In the later section of this chapter, experimental investigations are presented for the measurement of temperature, gaseous emissions, and radiation heat flux.

4.2 SPRAY STUDIES

Spray characterisation is carried out using a separate cold flow setup for measuring droplet mean diameter and cone angle for all operating conditions. A Danfoss make solid cone type pressure swirl nozzle 030F4904 is used for the present study. The nozzle is calibrated at two injection pressures of 9 bar and 12 bar to achieve two different fuel flow rates of 0.84 kg/hr (10 kW) and 4.67 kg/hr (20 kW). Details regarding the shadowgraphy are presented somewhere else [24]. Two important spray parameters—namely, Sauter mean diameter and cone angle—are obtained through spray studies. SMD is an important parameter in combustion studies wherein mixing is a crucial aspect. In the present study, flameless combustion is investigated and is achieved through increased levels of mixing and dilution. Also, the cone angle defines the spread of fuel cone in case of a solid-cone spray. High fuel cone spread can help achieve better mixing in some cases [24]. Tables 4.1 and 4.2 show the details regarding the fuel flow rates, cone angles, fuel flow velocities, SMD, and droplet number density (DND) [24].

TABLE 4.1
Summary of Atomiser Details, Injection Pressures, Flow Rates, and Flow Velocities, 9N1: N1 Atomiser at 9 bar Injection Pressure.

Sr No.	P_{inj} (bar)	$\dot{m}_f$ (kg/hr)	$Q_{thermal}$ (kW)	$\dot{Q}^{\cdot}$ (MW/m³)	V_f (m/s)
9N1	9	0.84	10	10	17.87
12N1	14	4.67	20	20	35.52

TABLE 4.2
Summary of Spray Characteristics with Kerosene Fuel.

P_{inj} (bar)	Cone Angle (°)	SMD (μm)	DND w.r.t. SMD (10⁴/cm³)
9	45	32.50	0.95
12	52	30.00	4.59

4.3 COMPUTATIONAL DETAILS

The flow turbulence is modelled using the seven-equation Reynolds stress model due to its ability to predict the high-intensity swirl flows [26] accurately. Different parameters such as swirl flow pattern, axial velocity, tangential velocity, and pressure drop, etc. are predicted correctly. P1 radiation model is used for the calculation of the radiative flux in the computational domain. It accurately predicts the radiation heat exchange in an industrial environment. Turbulent combustion is solved using a non-premixed droplet combustion model. Air-fuel chemistry is modelled using the chemical equilibrium approach along with probability density function (PDF) combustion model for capturing the turbulence-chemistry interaction. The PDF mixture containing 20 intermediate species is used to model the turbulence-chemistry interactions between the flow fluctuations and combustion reactions. In the present study, kerosene is specified as fuel with chemical formula $C_{12}H_{23}$ [20].

The modelling uses a mixture-fraction approach to calculate different combustion parameters such as species mass fraction, density, temperature, etc. using different chemistry models such as equilibrium, flamelet model, and non-equilibrium model. For turbulent flow, the average values of these fluctuating parameters are calculated. It is the turbulence-chemistry interaction which defines the accuracy of the instantaneous value prediction of these parameters. A probability density function (PDF) approach is used to calculate the average values of the fluctuations in the combustion parameters. A PDF table is generated in which time-averaged values of density, temperature, and mass fractions are derived from the mixture-fraction (f) values at all the points.

Droplet combustion is modelled using the discrete phase modelling, in which continuous interaction between gas and liquid phase is allowed. In this model, a fixed number of droplets are injected in the continuous gas flow after a fixed interval of gas-phase iterations. The gas and liquid phases are then allowed to interact and update the solution accordingly. Interaction of the gas-liquid phase and the solution updating goes on simultaneously until a steady-state is reached. The kerosene spray is defined as a separate injection in the form of a solid-cone spray with an appropriate Sauter mean diameter (SMD) and cone angle. Detailed information about the droplet size, cone angle, and fuel injection velocity is provided in this section. Appropriate values of spray parameters, (i) Sauter mean diameter (SMD), (ii) cone angle, and (iii) fuel flow rate is used for the solid-cone spray. These parameters are experimentally measured through detailed studies on injectors using shadowgraphy [24]. The shape of droplets is assumed to follow the spherical drag law.

There are different types of boundaries involved in the present work, including air inlet sections, combustor walls, and combustor outlet. Appropriate boundary conditions are needed for each of these sections to carry out the accurate numerical solutions. Mass flow inlet normal to boundary is applied at the different air inlet sections of the combustor. Turbulence for this boundary condition is specified using turbulent intensity and mean hydraulic diameter with appropriate values. Different air inlet temperatures are defined according to the requirement. Walls of combustor are modelled as isothermal walls, no-slip, and stationary. The combustor was not modelled for heat loss to the surroundings. In the present work, combustor walls were not insulated. Isothermal boundary condition based on average measured wall temperature was specified at the combustor walls for numerical simulations. However, it was challenging to calculate the heat transfer coefficient at every section of the combustor.

The exhaust of the combustor is specified as a pressure outlet with atmospheric pressure. All the air inlets and pressure outlets allow the reactants and products to escape; however, the walls are defined as reflective. The general-purpose CFD code, Fluent 16.1, offers four different algorithms for pressure-velocity coupling—namely, SIMPLE, SIMPLEC, PISO, and Coupled. For the present work, the SIMPLE algorithm is used to solve the pressure-velocity coupling for the calculation of the pressure from the discretised continuity equation. As the grid density is very high, the SIMPLE algorithm gives accurate prediction along with a reduced computation cost and running time. The second-order upwind discretisation scheme is used for precise prediction of the solution of various equations. For convergence check, a residual criterion for scaled residuals is kept as 1e-05 for all the parameters.

Figure 4.1a presents the dimensional details of the investigated combustor. All dimensions are in mm. The proposed combustor is optimised with the help of numerical calculations aimed at optimising the reactant dilution ratio, R_{dil}. It is a non-dimensional parameter used to quantify the amount of recirculation at a given plane. R_{dil} is defined in our previous studies [24]. It is to be noted that numerical calculations are aimed at optimising the reacting flow field. Based on the reacting flow field, the shape of the recirculation zone is observed, followed by a calculation of R_{dil}.

As seen in Figure 4.1a and b, the combustor is a stepped hollow cylindrical housing, with two rows of air injection holes drilled on the combustor wall. They are referred to as primary and secondary air injection holes. The proposed air injection method is aimed at achieving the required levels of dilution inside the combustor at increased thermal intensities. A bottom air swirler unit is attached at the inlet end of the combustor to control the flow of swirl air and the liquid fuel. The swirler has a 60° swirl angle for a better flow field [27]. A chamfered exit is provided at the outlet end of the combustor to help assist in recirculation of hot combustion products [22].

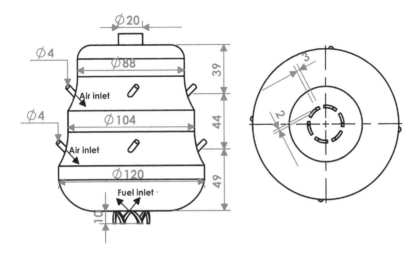

FIGURE 4.1 (a) Schematic of the computational domain of the flameless combustor and (b) pictorial view of the combustor.

More details regarding the number of holes, their diameters, and air distribution are shown in Table 4.3.

Combustor volume is 0.001 m³, and the thermal power is varied between 10 kW–20 kW, resulting in thermal intensities of 10–20 MW/m³. The computational simulations are carried out at equivalence ratio, $\phi = 4.0$, and the air inlet velocities are varied in the range of 20–99 m/s.

TABLE 4.3
Geometric Details of the Computational Domain.

Parameter	Primary Air	Secondary Air	Air Swirler
Nos.	4	4	6
Diameter (mm)	4	4	3×2
Air distribution (%)	50	20	30

4.3.1 AXIAL VELOCITY DISTRIBUTION

Figure 4.2 depicts the variation of axial velocities through a contour plot (Figure 4.2a) and a line plot (Figure 4.2b). The plots are obtained from the numerical calculations to help understand the existence of recirculation zones inside the combustor.

It is observed that recirculation zones are present throughout the combustor length and can be identified based on negative velocity contours. The proposed air injection scheme is primarily responsible for the formation of recirculation zones. However, the modified exit at the outlet end of the combustor further assists in inducing more recirculation. Increased recirculation level is critical to achieve flameless combustion [20] and is expected to have a significant effect on thermal and emissions profiles. Figure 4.2b shows the variation of axial velocity along the normalised radial coordinate for different axial locations. The distribution is symmetric about the axial direction; hence only half of the combustor length is presented here. As high as −21 m/s magnitude of axial velocity is calculated at 10 kW thermal input.

It is to be noted that recirculation zones are present from r/R = 0.1–4.0, showing the potential of this design for gas turbine applications. GT engines operate at high-thermal intensities with stringent combustor volume requirements. It is therefore difficult to achieve the required levels of dilution through aerodynamic means in a conventional can combustor. The proposed combustor design demonstrates the potential for its use in GT engines in the flameless mode of combustion.

As Figure 4.2a shows the presence of extended recirculation zones, it is equally important to understand the mechanism of their formation inside the running combustor. Figure 4.3 shows the schematic diagram of the mechanism of recirculation zone formation, which is explained on the basis of interactions of two separate air vortices. As seen from Figure 4.3, there are two different air vortices present inside the combustor. One is formed due to circumferential injection of primary and secondary air (presented as red colour curve). In contrast, the other is created due to bottom swirl air injection (shown as blue colour curve).

It is to be noted that both the vortices are of the same orientation (i.e. clockwise in the present study) while looking from the outlet end of the combustor. It is pointed out from the previous studies [27] that two counterrotating vortices negatively affect the formation of recirculation zones. When these air vortices interact with each other in the same orientation, a recirculation zone is formed along the edge of the inner vortex (blue colour). The shape of the modified exit in chamfered form also assists the recirculation and help extend their presence until the combustor outlet.

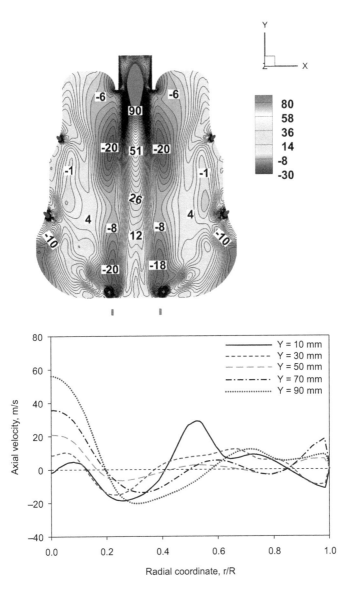

FIGURE 4.2 Axial velocity variation for reacting flow field for 10 MW/m³ at (a) XY plane of the combustor and (b) along the radial direction for different axial coordinates.

4.3.2 Reactant Dilution Ratio, R_{dil}

R_{dil} has been explained in the previous studies as a non-dimensional parameter to quantify the recirculation at a given plane for reacting flow [20]. R_{dil} is expressed as,

$$R_{dil} = \frac{\left| \dot{m}_{axial} \right| - (\dot{m}_{ox} + \dot{m}_f)}{(\dot{m}_{ox} + \dot{m}_f)} \tag{4.1}$$

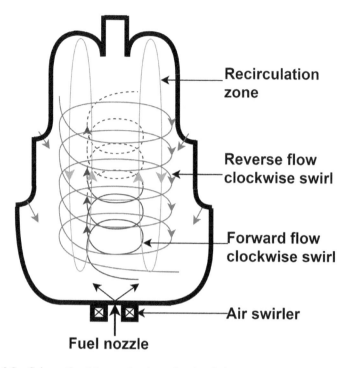

FIGURE 4.3   Schematic of the mechanism of recirculation zone formation.

Where $\dot{m}_{axial}$ is the total mass flow of hot combustion products at a given plane in the combustor, normal to the axis of the combustor and defined as $\dot{m}_{axial} = \iint \rho v_{axial} dx dz$, and here v_{axial} is the axial velocity. $\dot{m}_{ox} \, and \, \dot{m}_f$ are the oxidiser and fuel mass flow rates respectively. The density (ρ) and axial velocities of the exhaust products at each cell centre are obtained from the numerical simulations.

It is clear from Figure 4.4 that R_{dil} increases initially and attains a maximum at an axial location of 50 mm from the inlet end of the combustor and decreases further downstream. A maximum R_{dil} of 3.75 and 3.45 is calculated for thermal intensities of 10 MW/m³ and 20 MW/m³ respectively. As observed from Figure 4.4, R_{dil} decreases throughout the combustor length with increasing thermal input. It happens due to the increased number of fuel droplets at higher thermal input, as seen in Table 4.2. Mixing becomes more difficult with a large number of droplets leading to comparatively reduced levels of dilution. However, for the present study, the difference in maximum R_{dil} values at 10 MW/m³ and 20 MW/m³ is not significant. It is due to the reduction in droplet size for 20 MW/m³ in contrast to 10 MW/m³. A large number of smaller droplets are formed at higher thermal inputs. On one end, the finer spray helps improve the mixing while on the other end, a large number of droplets makes the mixing poor. The combined effect of these two phenomena leads to the peculiar behaviour of R_{dil}.

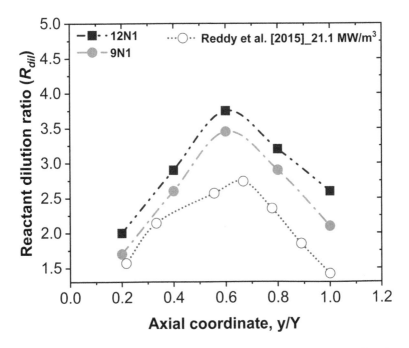

FIGURE 4.4   Variation of R_{dil} for varying thermal intensities along the axial direction.

4.4   EXPERIMENTAL DETAILS

Figure 4.5 shows the schematic diagram of the experimental setup used for the present study. The combustor is placed vertically on a test stand. Fuel supply unit is integrated along with the bottom air swirler at the inlet end of the combustor. Liquid kerosene is stored in a stainless-steel tank and supplied through a solid cone type pressure-swirl atomiser. Jackets are enveloped around the combustor wall to ensure the proper air distribution. Optical access inside the combustor is provided through a quartz window. It is attached by cutting through a 60° section of the combustor. The inner surface of the quartz window matches well with the inner surface of the combustor wall. The window is pressed and sealed to the wall of the combustor to avoid any leakage during the experimentation.

Airflow rate is maintained using Alborg make mass flow controllers of 1000 LPM, 750 LPM and 500 LPM capacities with an accuracy of ±4.5% of full scale. Air is heated up to a temperature of 800 K using a 36 kW electric heater. Omega make R type thermocouple of 0.25 mm wire diameter is used for temperature measurement. The response time of the thermocouple is 0.3 s, which is significantly longer than the turbulent time scale of ~3 ms. Therefore, the current measurements are taken over a time period of 60 s, and the average of the recorded data is reported. Measured temperatures are corrected for radiation losses, and it is observed that corrected values are deviated from the measured values by approximately 8%. Gaseous emissions are measured using TESTO 350 gas

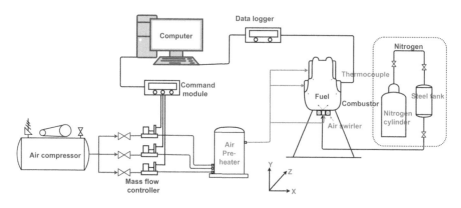

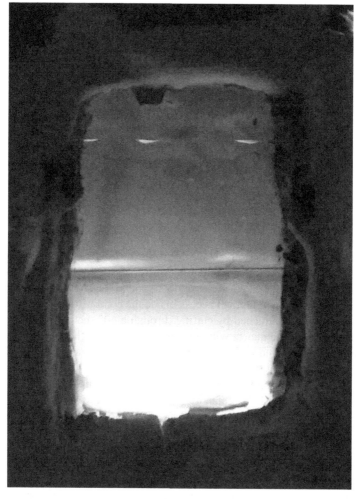

FIGURE 4.5 (a) Schematic diagram of the experimental set up and (b) pictorial view of the running combustor in flameless mode.

analyser. Details of the gas analyser are shown in Table 4.4. As a standard practice in the gas turbines, all the emissions are corrected to 15% O_2 level [2]. The method to convert the emission to 15% O_2 level is as follows:

$$X_{15\%,\,O_2} = X_i \frac{20.9 - 15}{20.9 - X_{O_2}} \tag{4.2}$$

where X_i is the measured emissions and X_{O_2} is the measured O_2 level. Measurements of radiative wall heat flux are carried out using MEDTHERM make Schmidt Boelter gauge. The responsivity of the gauge is 0.2237 mV/(kW/m²), and full-scale output is 13.42 mV at 60 kW/m² with an absorptance of 0.94.

For starting the combustor, fuel is supplied from the bottom, and the air is supplied from the swirler and primary holes. The combustor is allowed to run in the flame mode for 3–5 minutes. As the walls of the combustor are heated up well enough, the air supply is turned on from all the air injections, and fuel-air flow rates are adjusted to stoichiometric conditions. The increased momentum of hot air along with the proposed air injection scheme commences the mixing mechanism between hot air and combustion products. Following a short transition period, the combustor switches its mode into flameless combustion. As a result of that, combustion noise reduces significantly. The flame appearance changes from less visible to completely invisible flame. After the combustor switches into the flameless mode, measurements are carried out for temperature, gaseous and acoustic emissions.

TABLE 4.4
Measurement Range and Accuracies of Different Sensors.

Measurement	Measurement Range	Accuracy	Resolution
CO	0 to 10000 ppm	±5 ppm CO (0–199 ppm CO) ±5% of mv (200–2000 ppm to 2000 ppm CO) ±10% of mv (2001 to 10000 ppm CO)	1 ppm CO (0 to 10000 ppm CO)
NO	0 to 4000 ppm	±5 ppm NO (0 to 99 ppm NO) ±5% of mv (100 to 1999.9 ppm NO) ±10% of mv (2000 to 4000 ppm NO)	1 ppm NO (0 to 3000 ppm NO)
CO_2 (IR)	0 to 50 Vol. %	±0.3 Vol. % CO_2 + 1% of mv (0 to 25 Vol. % CO_2) ±0.5 Vol. % CO_2 + 4.5% of mv (>25 to 50 Vol. % CO_2)	0.01 Vol. % CO_2 (0 to 25 Vol. % CO_2) 0.1 Vol. % CO_2 (>25 Vol. % CO_2)
UHC	100 to 40,000	< 400 ppm (100 to 4000 ppm) < 10% of mv (> 4000 ppm)	10 ppm

4.4.1 Uncertainties in Measurements

Experimental uncertainties are presented for the measurements of air mass flow rate, temperature, gaseous and acoustic emissions. Air mass flow rate has a maximum uncertainty of ±15 LPM. Maximum uncertainties in the measured temperature are calculated as ±5 K; it includes both thermocouple accuracy ($4.5°C$ or 0.25%, whichever is max.) and the instrument accuracy ($\pm1\ °C$). Maximum uncertainties of ±1 dB are calculated in the acoustic level measurements. The estimated error for CO emissions is $\pm5\%$ of reading. NO_x emissions have an estimated error of ±2 ppm in the range of $0–99.9$ ppm and $\pm5\%$ of reading for the rest of the readings. The O_2 sensor has an error of $\pm0.2\,vol\%$ of reading.

4.5   RESULTS AND DISCUSSIONS

4.5.1 Temperature Measurement

Measurements are carried out at 60 mm downstream and the exit plane of the combustor. The recorded temperatures are compared with the predicted ones. Figure 4.6 shows the comparison between measurements and predictions for thermal intensities of 10 and 20 MW/m^3. It is observed that temperature is higher at the 60 mm location, in contrast to the exit plane due to increased combustion intensity in the near nozzle region [24]. For a specific thermal input, the maximum temperature is recorded at the centre and decreases towards the combustor wall.

It is observed that the numerical model slightly underpredicts the measurements in the central zone of the combustor. Selection of an isothermal wall boundary condition could justify this behaviour as it was not in the scope of the present work to calculate the heat transfer coefficient at different sections of the combustor. Maximum temperatures of 1580 K and 1611 K are reported for thermal intensities of 10 MW/m^3 and 20 MW/m^3 respectively. Corresponding wall temperatures are recorded as 1299 K and 1352 K, hence indicating the flatness of the thermal field as the differences between the centreline and wall temperature, $\varnothing T_W$, are 281 K and 259 K for the cases of 10 MW/m^3 and 20 MW/m^3 respectively. It also shows that the thermal field becomes comparatively more uniform for higher heat intensity. It can be seen in Figure 4.6b that the temperature profile looks almost flat at the exit plane, an important advantage of this design as far as the life span of the turbine blades is considered. Maximum centreline temperature at combustor exit is recorded as 1440 K and reduces to 1352 K at the combustor wall, resulting in a thermal gradient of ~88 K.

4.5.2 Emission Analysis

Measurements of CO and NO_x emissions are carried out for the representative cases of 10 MW/m^3 and 20 MW/m^3 respectively with varying equivalence ratios. All the emissions are corrected to 15% O_2 level as a standard practice in gas turbine applications. CO shows an increasing trend with increasing thermal input. Also, lean fuel-mixture appears to produce more CO emissions compared to a stoichiometric

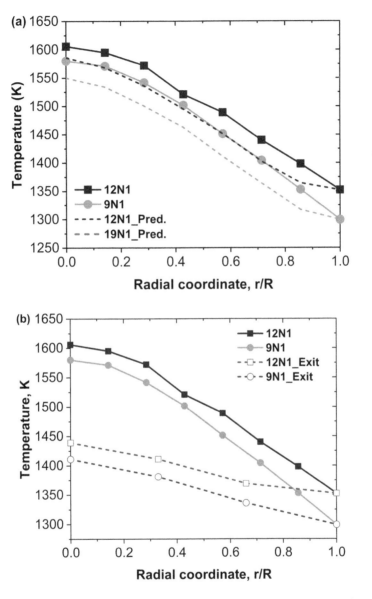

FIGURE 4.6 Variation of temperature in the radial direction; (a) comparison of predicted and measured temperatures at 60 mm downstream and (b) comparison of recorded temperatures at 60 mm downstream and at the exit of the combustor.

mixture. It is concluded that although increasing thermal input results in increased temperatures, however, it reduces the residence time too. Both these factors (i.e. increased temperature and reduced residence times) oppositely affect the CO emissions. The increased temperature speeds up the CO oxidation reaction and reduced

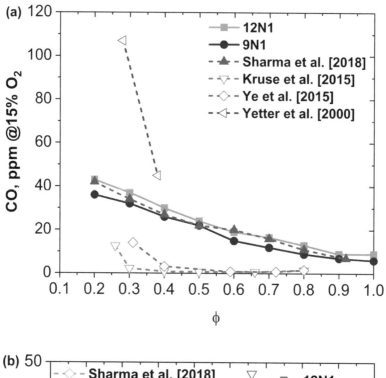

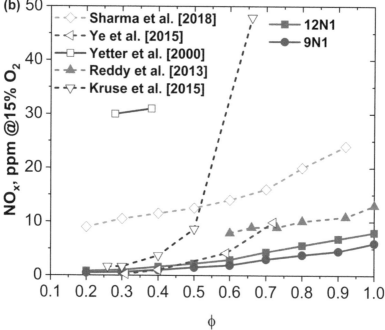

FIGURE 4.7 Variation of (a) CO emissions and (b) NO_x emissions for varying thermal intensities of 10–20 MW/m³.

residence times restrict the $CO \rightarrow CO_2$ conversion. A combined effect of these two mechanisms results in this peculiar behaviour of CO emissions (see Figure 4.7a). The recorded emissions are compared with the literature data on flameless/MILD combustion. It is observed that CO are well within limits in comparison to conventional combustion [28].

Figure 4.7b shows the variation of NO_x emissions for all the operating conditions of thermal intensities and equivalence ratios. NO_x stays below 5 ppm and 8 ppm for 10 MW/m^3 and 20 MW/m^3 respectively. Such low levels of NO_x emission despite high thermal intensities attribute to the increased levels of mixing and dilution, achieved through the novel geometric features of the combustor. The NO_x approaches zero emissions in the ϕ range of 0.2–0.4 for both thermal intensities. NO_x is observed to increase with increasing thermal intensity due to comparatively higher temperature inside the combustor. Also, low overall NO_x is produced for lean operating conditions in comparison to stoichiometric conditions due to both reduced residence times and low overall temperatures.

4.5.3 RADIATION HEAT FLUX

Radiation heat flux has been presented as a parameter to depict the attainment of flameless combustion in the industrial furnace [29]. For the present study, the flux is measured at 60 mm downstream by inserting the MEDTHERM make Schmidt Boelter gauge such that the plane of the gauge matches well with the inner surface of the combustor wall. Figure 4.8 shows the variation of heat flux for a period of 20 s during both conventional and flameless modes at 20 MW/m^3.

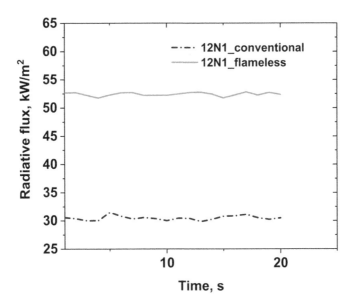

FIGURE 4.8 Instantaneous variation of radiation heat flux in the conventional and flameless mode.

Initially, an average heat flux of ~30 kW/m² is recorded during the conventional mode attributing to the presence of a flame in the central zone of the combustor. As the combustor starts operating at stoichiometric condition, the combustion switches into the flameless mode and heat flux shoots up to an average value of ~52 kW/m², thus increasing the flatness of the thermal field. High and flat profiles of radiation heat flux attribute to flameless combustion conditions [29].

4.6 CONCLUSIONS

The present study aims at the development and testing of a can-type GT combustor operating in flameless mode. The combustor geometry is optimised based on numerical calculations followed by experiments after that. Significant concluding remarks are as follows:

1. Computations show the presence of large recirculation zones inside the optimised combustor for a complete range of thermal intensities. Modifying the exit of the combustor further aids the recirculation of hot combustion products and helps extends the recirculation zone until the combustor exit plane.
2. Variation of R_{dil} shows improvement over the previous study on flameless combustion and a maximum R_{dil} value of 3.75 is calculated at 50 mm downstream for 10 MW/m³ thermal intensity.
3. Temperature distribution shows uniformity at 50 mm downstream, and a nearly flat thermal field is observed at the exit plane of the combustor.
4. Combustor emits nearly zero NO_x emissions in the range of $\phi = 0.2–0.4$, the lean operational range in which GT combustor usually operate.
5. The developed combustor shows a promising candidature by producing ultra-low emissions at GT combustor relevant conditions (~20 MW/m³).

REFERENCES

[1] A. Flightpath, 2050—Europe's vision for aviation, Advisory Council for Aeronautics Research in Europe, 2011.
[2] S. Kruse, B. Kerschgens, L. Berger, E. Varea, H. Pitsch, Experimental and numerical study of MILD combustion for gas turbine applications, *Appl. Energy* 148 (2015) 456–465.
[3] J. Ye, P.R. Medwell, E. Varea, S. Kruse, B.B. Dally, H.G. Pitsch, An experimental study on MILD combustion of prevaporised liquid fuels, *Appl. Energy* 151 (2015) 93–101.
[4] V.K. Arghode, A.K. Gupta, Development of high intensity CDC combustor for gas turbine engines, *Applied Energy* 88 (2011) 963–973.
[5] V.K. Arghode, A.K. Gupta, K.M. Bryden, High intensity colorless distributed combustion for ultra low emissions and enhanced performance, *Appl. Energy* 92 (2012) 822–830.
[6] A.E. Khalil, A.K. Gupta, On the flame-flow interaction under distributed combustion conditions, *Fuel* 182 (2016) 17–26.
[7] A.E. Khalil, A.K. Gupta, Fostering distributed combustion in a swirl burner using prevaporized liquid fuels, *Applied Energy* 211 (2018) 513–522.

[8] G. Sorrentino, P. Sabia, M. De Joannon, R. Ragucci, A. Cavaliere, U. Göktolga, J. Van Oijen, P. De Goey, Development of a novel cyclonic flow combustion chamber for achieving MILD/flameless combustion, *Energy Procedia* 66 (2015) 141–144.

[9] M. de Joannon, P. Sabia, G. Sorrentino, P. Bozza, R. Ragucci, Small size burner combustion stabilization by means of strong cyclonic recirculation, *Proceedings of the Combustion Institute* 36 (2017) 3361–3369.

[10] R. Lückerath, W. Meier, M. Aigner, FLOX® combustion at high pressure with different fuel compositions, *Journal of Engineering for Gas Turbines and Power* 130 (2008).

[11] O. Lammel, H. Schütz, G. Schmitz, R. Lückerath, M. Stöhr, B. Noll, M. Aigner, M. Hase, W. Krebs, FLOX® combustion at high power density and high flame temperatures, *Journal of Engineering for Gas Turbines and Power* 132 (2010).

[12] R. Sadanandan, R. Lückerath, W. Meier, C. Wahl, Flame characteristics and emissions in flameless combustion under gas turbine relevant conditions, *Journal of Propulsion and Power* 27 (2011) 970–980.

[13] H. Schütz, O. Lammel, G. Schmitz, T. Rödiger, M. Aigner. EZEE®: A high power density modulating FLOX® combustor. In: editor^editors. Turbo Expo: Power for Land, Sea, and Air; 2012: American Society of Mechanical Engineers. pp. 701–712.

[14] Y. Levy, V. Sherbaum, P. Arfi, Basic thermodynamics of FLOXCOM, the low-NOx gas turbines adiabatic combustor, *Appl. Therm. Eng.* 24 (2004) 1593–1605.

[15] M. Melo, J. Sousa, M. Costa, Y. Levy, Experimental investigation of a novel combustor model for gas turbines, *J. Prop. Power* 25 (2009) 609–617.

[16] M. Melo, J. Sousa, M. Costa, Y. Levy, Flow and combustion characteristics of a low-NOx combustor model for gas turbines, *J. Prop. Power* 27 (2011) 1212–1217.

[17] Y. Levy, A. Rao, V. Sherbaum. Chemical kinetic and thermodynamics of flameless combustion methodology for gas turbine combustors. In: editor^editors. 43rd AIAA/ASME/SAE/ASEE Joint Propulsion Conference & Exhibit; 2007. p. 5629.

[18] A.G. Rao, F. Yin, J.P. van Buijtenen, A hybrid engine concept for multi-fuel blended wing body, *Aircraft Engineering and Aerospace Technology: An International Journal* (2014).

[19] Y. Levy, V. Erenburg, V. Sherbaum, I. Gaissinski. Flameless oxidation combustor development for a sequential combustion hybrid turbofan engine. In: editor^editors. Turbo Expo: Power for Land, Sea, and Air; 2016: American Society of Mechanical Engineers. p. V04BT04A055.

[20] V.M. Reddy, D. Sawant, D. Trivedi, S. Kumar, Studies on a liquid fuel based two stage flameless combustor, *Proc. Combust. Inst.* 34 (2013) 3319–3326.

[21] V.M. Reddy, D. Trivedi, D. Sawant, S. Kumar, Investigations on emission characteristics of liquid fuels in a swirl combustor, *Combust. Sci. Technol.* 187 (2015) 469–488.

[22] V.M. Reddy, A. Katoch, W.L. Roberts, S. Kumar, Experimental and numerical analysis for high intensity swirl based ultra-low emission flameless combustor operating with liquid fuels, *Proc. Combust. Inst.* 35 (2015) 3581–3589.

[23] V.M. Reddy, P. Biswas, P. Garg, S. Kumar, Combustion characteristics of biodiesel fuel in high recirculation conditions, *Fuel Process. Technol.* 118 (2014) 310–317.

[24] S. Sharma, H. Pingulkar, A. Chowdhury, S. Kumar, A new emission reduction approach in MILD combustion through asymmetric fuel injection, *Combust. Flame* 193 (2018) 61–75.

[25] S. Sharma, R. Kumar, A. Chowdhury, Y. Yoon, S. Kumar, On the effect of spray parameters on CO and NOx emissions in a liquid fuel fired flameless combustor, *Fuel* 199 (2017) 229–238.

[26] B. Danon, W. De Jong, D. Roekaerts, Experimental and numerical investigation of a FLOX combustor firing low calorific value gases, *Combustion Science and Technology* 182 (2010) 1261–1278.

[27] S. Sharma, A. Chowdhury, S. Kumar, A novel air injection scheme to achieve MILD combustion in a can-type gas turbine combustor, *Energy* 194 (2020) 116819.

[28] R.A. Yetter, I. Glassman, H.C. Gabler, Asymmetric whirl combustion: A new low NOx approach, *Proc. Combust. Inst.* 28 (2000) 1265–1272.

[29] R. Weber, J.P. Smart, W. vd Kamp, On the (MILD) combustion of gaseous, liquid, and solid fuels in high temperature preheated air, *Proc. Combust. Inst.* 30 (2005) 2623–2629.

5 Combustion Aspects of Non-Conventional Reciprocating Internal Combustion Engines

Vijayashree and V. Ganesan

CONTENTS

DOI: 10.1201/9781003049005-5

5.1   INTRODUCTION

Reciprocating internal combustion engines can be classified as Spark Ignition (SI) and Compression Ignition (CI) engines. These two engines were invented during the last quarter of 19th century. In the latter half of 20th century they underwent tremendous design to reach their present-day astronomical level. They have established themselves as the best power train for the mobility sector. As people were greatly dependent upon horses in 19th century for their mobility, today we are completely dependent on these two engines for our mobility. It is interesting to note that when the first horseless carriage—viz., automobile—came on the American roads in 1895 the following statement was made.

> This discovery begins a new era in the history of civilization. It may someday be more revolutionary in the development of human society than the invention of wheel, reported the Joint Committee of the US Congress in 1895 when the first horseless carriage, i.e. the automobile rolled on the road. At the same time, it said, never in history, has society been confronted with a power so full of potential danger and the same time so full of promise for the future of man and for the peace of the world.

This statement has become 100% true today. It is an irony that humankind wanted to get rid of horses because of their semi-solid and liquid pollution and today we are in the same situation in wanting to get rid of these two engines as they are polluting the environment with gaseous pollutants. It is interesting to note that early in the 20th century almost all people owned horses except a few rich people who owned a car. But in the early 21st century we see that almost all people own a car but only a few rich people own horses.

Since almost all people own a car, pollution has become the greatest problem nowadays. Therefore, humankind is at a crossroads again to tackle the problem of environmental pollution. Replacing all the existing vehicles is not going to be simple as we think. In the authors' opinion it will take a minimum of 80 years to completely get rid of them. Therefore, we have to live with these two engines in this century.

These engines have been developed over a period of 100 years. It was the practice to use carburetion in an SI engine and mechanical injection in a CI engine. Over the years quite a lot of research has been carried out to improve the performance of these engines and because of this a number of new developments have taken place. Of all the processes in a reciprocating internal combustion engine, the combustion process is the most important one. In this chapter, we will look into four of these modern non-conventional engines and their combustion process in a comprehensive manner. The engines that will be discussed are:

- Common Rail Direct Injection (CRDI) engine.
- Gasoline Direct Injection (GDI) engine.
- Homogeneous Charge Compression Ignition (HCCI) engine.
- Stratified Charge (SC) engine.

5.2   HISTORY OF THE COMMON RAIL DIRECT INJECTION (CRDI) ENGINE

The very first development of CRDI was carried out by Robert Huber of Switzerland. Dr. Marco Ganser at the Swiss Federal Institute of Technology in Zurich developed this technology further in 1995. Japan has the credit of successfully developing the first working CRDI engine in the mid-1990s. Dr. Shohei Itoh and Masahiko Miyaki of the Denso Corporation claimed the first commercial high-pressure common rail system in 1995. The modern CRDI system is controlled electronically unlike the old systems which were controlled mechanically. The electronic system was extensively prototyped in the 1990s with collaboration between Magneti Marelli, Centro Ricerche Fiat, and Elasis. After considerable research and development by the Fiat Group, the design was acquired by the German company Robert Bosch GmbH for completion of development and refinement for mass production. A large number of present day diesel engines employ Common Rail Direct Injection systems. It is mainly because of the flexibility they offer while satisfying the ever-increasing stringent emission control standards.

5.3   COMMON RAIL DIRECT INJECTION

Gasoline and diesel are being used in conventional Spark Ignition and Compression Ignition engines respectively, as fuels for energy input for developing power. Diesel, being highly viscous, has slow evaporation characteristics. When injected using a conventional injection system the particles are heavier and atomization becomes a problem. Because of this, soot particles are produced during combustion, and this causes pollution of the environment. Further, performance gets degraded. In order to take care of all these problems, over the years, high-pressure injection systems have been developed for both SI and CI engines. Engines that use high-pressure injection technology for better combustion are called Common Rail Direct Injection (CRDI) engines. This technology is more appropriate for diesel engines, as injection pressures up to 1500 bar are employed. Conventional diesel engines employ around 300 bar for each injection. It is to be noted that in CRDI engines, the pressure is built up independently in the fuel line and does not depend upon the injection sequence. Sensor technology for individual cylinders is employed in Common Rail Direct Injection systems so as to evaluate real-time combustion data.

The common rail, usually called an accumulator, is placed above the cylinders. Fuel injection takes place at a constant pressure of 1500 bar which is controlled by an Electronic Control Unit (ECU). The ECU is the heart of the system, which operates high-speed solenoid valves. The quantity of fuel to be injected is regulated by an electronic Engine Management System (EMS). Depending upon the need and load on the engine, EMS controls the injection timing and the amount of fuel for each cylinder. In other words, compared to the conventional diesel engines, pressure generation and fuel injection are separately handled by EMS. Since the fuel is injected at 1500 bar the fuel particles are finely atomized and dispersed in each cylinder effectively. This is considered a great advantage for improved combustion.

In conventional engines, metering of fuel accurately for each cylinder is a problem. As this is taken care of by ECU in CRDI engines, better atomization, improved combustion and reduced pollution are achievable. It is reported in the literature that fuel economy improves by one-third (i.e. 30%) in CRDI engines. A substantial reduction in noise, vibration and harshness (NVH) could also be achieved because of a better synchronized timing of fuel injection. This technology paved the way for Gasoline Direct Injection in SI engines which will be discussed in the next section.

5.3.1 THE WORKING PRINCIPLE

A typical CRDI system is shown in Figure 5.1. The technology is called Common Rail Direct Injection, as the fuel is injected into each cylinder from a high-pressure common rail. The highly pressurized fuel (1500 bar) is kept in the common rail and connected to the fuel injectors of each cylinder. As can be seen in the figure, fuel from the fuel tank passes through a fuel filter to a low-pressure pump where the pressure is raised to a reasonably high value. Then the fuel goes to a high-pressure pump and a pressure regulator valve controls the pressure. In the modern CRDI system, the outlet pressure can be as high as 1800 bar. The highly pressurized fuel goes to an accumulator, which is called common rail. The pressure in the common rail is monitored by the rail pressure sensor and the fuel temperature is monitored by a temperature sensor.

In CRDI system, the fuel is distributed to each cylinder at a pressure of 1500 bar from the reservoir of fuel, which is nothing but the common rail. This simplifies the purpose of the high-pressure pump because the system has to maintain only the designed pressure either mechanically or electronically. When the fuel injectors are electrically activated a hydraulic valve (consisting of a nozzle and plunger) is mechanically or hydraulically opened and fuel is sprayed into the cylinders at the desired pressure. The fuel injectors are ECU-controlled.

5.3.2 THE INJECTOR

A typical fuel injector used in CRDI is shown in Figure 5.2a. The injector, which is nothing but a valve, is controlled electronically and thereby capable of opening and closing the required number of times per second. The pressurized fuel is distributed through a tiny nozzle to each individual cylinder at the appropriate time. For this, the injector is energized and thereby an electromagnet moves a plunger that opens the

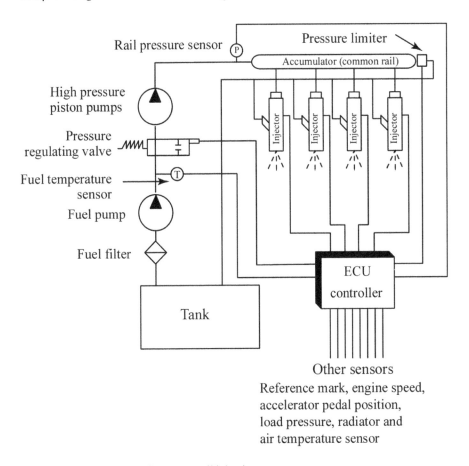

FIGURE 5.1 Schematic of common rail injection system.

Source: Reproduced from the book *Internal Combustion Engine* by V. Ganesan, fourth edition, with permission from McGraw Hill Education New Delhi.

valve. Each injector is built with a nozzle, hydraulic intensifier and digital electronic valve. The digital electronic valve is nothing but a rapid-acting solenoid valve that adjusts both the injection timing and the quantity of fuel to be injected.

A microcomputer controls each valve's opening and closing sequence. The nozzle is designed to inject the fuel as fine a mist as possible so that it can evaporate extremely fast and burn at a rapid rate. The amount of fuel supplied to the engine depends upon the amount of time the fuel injector stays open. This is called the pulse width, and it is controlled by the ECU. Since, the injectors are electrically actuated, the injection pressure at the start and end of injection is very close to the pressure in the accumulator (rail), thus producing a square injection rate. The high-pressure injectors are available with different nozzles for different spray configurations. A swirler nozzle produces a cone-shaped spray and a slit nozzle for a fan-shaped spray. The details can be seen in Figure 5.2b.

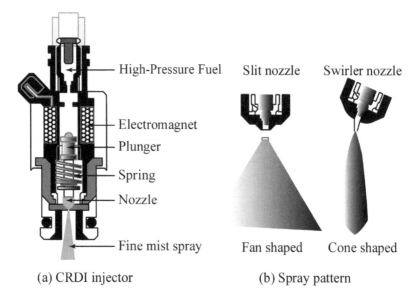

(a) CRDI injector   (b) Spray pattern

FIGURE 5.2   A typical CRDI injector.

Source: Reproduced from the book *Internal Combustion Engine* by V. Ganesan, fourth edition, with permission from McGraw Hill Education New Delhi.

5.3.3 SENSORS

The use of microprocessors and sensors helps to control the combustion process accurately; thereby better and most efficient use of the fuel is possible. The following input sensors are used in CRDI engines.

1. Crank position sensor.
2. Lambda sensor.
3. Pressure sensor.
4. Throttle position sensor.

Further, two other sensors for monitoring temperature and pressure are also employed. Because of the electronic management, power, fuel-economy and performance of the engine are very much improved. The Electronic Control Unit (ECU) and electronic display unit (EDU) are mainly responsible for adjusting injection pressure precisely from the data input from various sensors and display all important parameters.

5.3.4 ELECTRONIC CONTROL UNIT (ECU)

Solenoid and/or piezoelectric valves are employed for the precise control of fuel injection time, quantity and the pressure. As mentioned earlier, because of the improved fuel atomization due to high-pressure injection, better combustion and lower engine noise are achieved. Further, the engine's Electronic Control Unit can make a small quantity "pilot" injection just before the main injection event. Moreover, ECU can take into consideration the quality of fuel and cold starting condition in optimizing injection timing and quantity. Because of the precise timing by ECU, a CRDI system can monitor the post-combustion process also. During this process, a small amount of fuel is injected during the expansion stroke thus creating small-scale combustion after the normal combustion is completed. This greatly reduces the unburnt particles and thereby the exhaust gas temperature is increased which will be helpful for supercharging. Further, the pre-heat time of the catalytic converter gets very much reduced. This paves way for reduced emission during the post-combustion process.

In a CRDI system, the drive torque fluctuation and high-pressure pulsation in the lines are minimal. It is due to ECU, which takes care of the exact quantity of fuel supplied by the pump. The modern CRDI system has provision for four or five precise injections per stroke. Thus, a CRDI system can be considered as an intelligent system controlling a diesel engine employing modern electronics.

In a conventional diesel engine, mechanical components control the interval between injections and the fuel quantity, but in a CRDI system the time interval and timing etc. are all controlled by a central computer or microprocessor-based Electronic Control Unit. In the current first-generation design, the pipe withstands pressures as high as 1500 bar. Precise timing reduces the characteristic "diesel knock" common to all diesel engines, direct injection or not. To run a CRDI system effectively and efficiently, the microprocessor must be programmed beforehand to ably assist input from multiple sensors to calculate the precise amount of the diesel and the timing. Thus, the CRDI control system delivers the right amount of diesel at the right time to allow the best possible output with the least emissions and least possible wastage of fuel.

5.3.5 Microcomputer

In a modern CRDI system powerful microcomputers are employed to control the direct injection motors linked via CAN (Controller Area Network) database to other control devices on board to exchange data. The engine's electrical control is the heart of the common rail system. Regulation of injection pressure and control of the solenoid valves for each cylinder is handled by the microcomputer. Therefore, without a microcomputer a CRDI system would be unthinkable. This electronic engine management network is the critical element of the common rail system because only the speed and spontaneity of electronics can ensure immediate pressure injection adjustment and cylinder-specific control of the solenoid valve injectors.

The microcomputer regulates the time duration for the valves to stay open for the amount of fuel to be injected which depends upon the operating conditions and the required output. When it shuts the solenoid valves, fuel injection stops spontaneously. With the state-of-the-art common rail direct fuel injection, an ideal compromise can be attained between economy, long life, ride comfort and torque.

5.3.6   THE BENEFITS OF COMMON RAIL INJECTION

As already stated, noise, vibration and harshness (NVH) will be very much under control with CRDI mainly due to the timing flexibility. Therefore, the engine runs quieter and smoother. Because of very high injection pressure a finer spray is produced that burns more efficiently.

Better combustion efficiency means meeting emission standards. Less fuel is wasted as soot or particulates in the exhaust and deposits in the engine. A cleaner running engine is good for the environment and improves the long-term durability and reliability of the engine.

Another important to note is this technology generates an ideal swirl in the combustion chamber. Coupled with the common rail injectors' superior fuel-spray pattern and optimized piston head design, the air-fuel mixture forms a perfect vertical vortex. This results in uniform combustion and greatly reduces NO_x (nitrogen oxide) emissions by more than 50% over the current generation of diesel engines.

The new common rail engine (in addition to other improvements) cuts fuel consumption by 30%, doubles torque at low engine speeds and increases power by 25%. It also brings a significant reduction in the noise and vibrations of conventional diesel engines. In emission, greenhouse gases (CO_2) are reduced by 30%. At a constant level of NO_x, carbon monoxide (CO) emissions are reduced by 40%, unburnt hydrocarbons (HC) by 50% and particulate emissions by 60%.

5.3.7   ADVANTAGES OF CRDI SYSTEMS

By introduction of CRDI a lot of advantages can be achieved. Some of them are:

*   Improved power, increased fuel efficiency and reduced noise.
*   Fewer emissions and reduced particulates in the exhaust.
*   Precise injection timing, very high injection pressure and better stability.
*   Better pulverization of fuel.
*   Increased combustion quality due to pilot and post injection.
*   Doubling the torque at lower speeds.

The main disadvantage is that this technology increases the initial cost and the maintenance of the engine.

5.4   GASOLINE DIRECT INJECTION ENGINE

Until the late 20th century, carburettors were employed for fuel control in automobiles; even in 2020, carburettors are still used in many machines, such as two-wheelers, lawnmowers and chainsaws. With improvements in automobile design and development, the carburettor design became more and more complicated in order to take care of all the operating requirements. To handle some of these tasks, carburettors had five different circuits making it more and more complicated.

*   *Main circuit:* provides appropriate quantity of fuel for fuel-efficient cruising.

- *Idle circuit*: provides just enough fuel to keep the engine running at idling speed.
- *Accelerator pump:* provides an extra amount of fuel for acceleration.
- *Power enrichment circuit:* Provides additional fuel for climbing a hill.
- *Choke*: Provides extra rich mixture during cold start.

As emission norms become more and more stringent day by day carburettors could not meet stricter emissions norms. Therefore, catalytic converters were introduced. For better results, air-fuel ratios are to be controlled very precisely and this is not feasible with a mechanical device like a carburettor. Then electrical carburettors were introduced for a brief period and were found to be even more complicated than the mechanical ones.

The history of fuel feed systems for SI engines is shown in Figure 5.3. Throttle body fuel injection systems were introduced between 1980 and 1985. They were called single point or central fuel injection systems. They were electrically controlling the fuel-injector valves into the throttle body. In those years, as automobile manufacturers did not want any drastic change in the design of their engines, the cost was also a major consideration.

Over a period, new engines were designed and developed. Throttle body fuel injection was replaced by multi-port fuel injection between 1995–2005. They were also port injection, multi-point fuel injection or sequential fuel injection engines. With each cylinder having a fuel injector, the right quantity of fuel could be sprayed at the intake valve. This was better than throttle body injection.

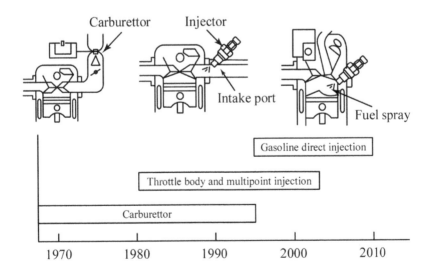

FIGURE 5.3   History of carburettors.

Source: Reproduced from the book *Internal Combustion Engine* by V. Ganesan, fourth edition, with permission from McGraw Hill Education New Delhi.

Between 1995 and 2005, Gasoline Direct Injection was introduced into four-wheelers. After 2005, all major automobile manufacturers switched over to Gasoline Direct Injection (GDI) technology. This is similar to CRDI in diesel engine. The gasoline is pressurized, and injected via a common rail fuel line directly into the combustion chamber of each cylinder. This is different from conventional multi-point fuel injection (MPFI) in which injection takes place in the intake manifold, or cylinder port. In some applications, Gasoline Direct Injection enables stratified fuel charge (ultra-lean burn) combustion for improved fuel efficiency, and reduced emission levels at low load.

5.4.1 PRINCIPLE OF OPERATION

The major advantage with GDI technology is better fuel economy and improved power output and lesser emissions. These could be achieved because of metering of exact fuel quantity and precise injection timing for various load and speed conditions by ECU. Losses due to throttling and pumping could be very well minimized, thereby, improved efficiency is achieved. Regulated and accurate injection timing enables speed control through the Engine Management System (EMS). However, achieving this control through EMS requires a large memory in ECU. To achieve better performance, control and driveability this is a must. The Engine Management System continually chooses among three combustion modes—viz., ultra-lean, stoichiometric and full power output—based on the air-fuel ratio. For most of the hydrocarbon fuels stoichiometric air-fuel ratio is close to 15:1. But, ultra-lean mode can be 400% leaner (60:1) or even higher in some engines, for a very limited period. These mixtures are very much leaner than in a conventional carburetted engine and reduce fuel consumption quite considerably. Let us look into the details in the following paragraphs.

Ultra-lean burn mode: In this mode the engine may run at very low load, at constant or reduced road speeds and probably no acceleration is required. During this operating condition the fuel injection is almost nil in the intake stroke and the fuel injection takes place only at the end of the compression stroke. Because of this, a small amount of stoichiometric air-fuel mixture is optimally placed near the spark plug. This stratified charge is surrounded mostly by air and this is the secret for its success. This keeps the fuel and the flame away from the cylinder walls by which the lowest emissions and heat losses are achieved. The combustion takes place in a toroidal (donut-shaped) cavity on the piston head. The cavity is offset towards the fuel-injector side. This technique enables the use of ultra-lean mixtures that otherwise would be impossible with carburettors or conventional fuel injection.

Stoichiometric mode: This comes into operation under medium load conditions. In this mode fuel is injected during the intake stroke. This creates a homogenous fuel-air mixture in the cylinder close to a stoichiometric condition bringing the best combustion results. This causes minimum harmful exhaust emission, further cleaned by the catalytic converter.

Full power mode: It is used for rapid acceleration and during heavy loads (climbing a hill). The air-fuel mixture is homogenous and the ratio is slightly richer than stoichiometric. This can help in preventing knock. In this mode fuel is injected during the intake stroke.

Direct injection may also be accompanied by other engine technologies such as variable valve timing (VVT) and tuned/multi path or variable length intake manifolding (VLIM, or VIM). Further, water injection or more commonly employed exhaust gas recirculation (EGR) may help in reducing the high nitrogen oxides (NO_x) emissions that can result from burning ultra-lean mixtures.

As entire operation is controlled by electronics multiple injections are possible. After the first fuel charge has been ignited, it is possible to add fuel during the expansion stroke which will culminate into better power and economy. A direct injection engine, where the injector injects directly into the cylinder, is limited to the suction stroke of the piston. As the engine speed increases, the time available to inject fuel decreases. Newer FSI systems that have sufficient fuel pressure to inject even late in the compression phase do not suffer from this deficiency. It should be kept in mind that the octane number of the fuel has a significant effect and therefore choosing fuel with an appropriate octane number is important; otherwise it can cause exhaust valve erosion.

5.4.2 Advantages of Gasoline Direct Injection

The following are some of the advantages of direct injection:

Power Output—Except some high end modified vehicles, Gasoline Direct Injection provides a much better power output and performance than a standard carburetted engine.

In a Gasoline Direct Injection system, as the fuel is pressurized to a higher pressure and dispersed into the cylinder, it has room to expand thereby causing temperature drop within the cylinder. This causes a cooling effect in the cylinder in which injection occurs. Therefore, there is a room for a more aggressive timing profile, for both fuel and injection, providing better power.

Fuel Efficiency—As a vehicle's ECU controls fuel injection, management of fuel consumption is much better than with a carburettor, resulting in better fuel efficiency.

Emissions Performance—Better fuel efficiency results in better emissions performance. GDI engines produce far fewer carbon-based emissions than vehicles with carburettors. This is due to the engine using the just needed fuel, which reduces emissions because of exact metering.

Alternate Fuel Accommodation—GDI engines are better suited for handling alternative fuels, and fuels with additives. These help keep your car engine clean. A good example of this is e-85, also called flex fuel. It is significantly cheaper than gasoline but may not run at the same efficiency as gasoline.

Drivability—GDI engines run much smoother. This is because there's better fuel flow management and more consistency. The engine only uses the fuel that it needs to get to the required speed, which is great for saving fuel. When running at a constant speed, the air to fuel ratio is leaner.

Diagnostic Capability—As the engine is computer regulated in GDI, diagnosing problems is easy with a simple computer diagnostic test. Instead of tearing the engine apart to figure out the problem, a simple diagnostic tool can figure it out. This is due to the precise nature of direct injection. Therefore, this can save time and money at the repair shop.

While great, the power of the GDI engine is a double-edged sword. While you get many benefits from Gasoline Direct Injection, carbon deposits that build up can steal the entire positive attributes associated with this type of engine. Like CRDI in diesel engines, the main disadvantage is that this technology increases the initial cost and the maintenance of the engine.

5.5   HOMOGENEOUS CHARGE COMPRESSION IGNITION (HCCI) ENGINE

Homogeneous Charge Compression Ignition (HCCI) is a form of internal combustion in which a homogeneously mixed charge is compressed and ignited (HCCI) through auto-ignition. The sensible chemical energy of the fuel is released by exothermic reaction and converted to work.

HCCI combines the two forms of combustion in SI and CI engines, viz.,

- Homogeneous charge spark ignition.
- Heterogeneous charge compression ignition.

Because of higher compression ratio, the entire mixture reaches its self-ignition temperature and reacts spontaneously. There is an important point to note—viz., occurrence of spontaneous ignition at multiple points without any spark plug being used. This is a challenge for the designers, to design an appropriate combustion chamber. However, advanced electronics and our understanding of the ignition process help us in overcoming this challenge.

Conventional SI engines have good full load performance whereas the CI engines have better part load performance. The Homogeneously Charged Compression Ignition concept, if properly implemented, can achieve gasoline engine-like emissions along with diesel engine-like efficiency. The NO_x emissions from HCCI engines are very low and therefore no after treatment is necessary. However, the HC and CO emissions are still high (due to lower peak temperatures), as in SI engines. Therefore, exhaust gas treatment is a must to meet the stringent emission regulations.

As already mentioned, controlling combustion is a challenge. In order to face this challenge a duel fuel concept such as gasoline and diesel are tried to solve some of the difficulties of controlling HCCI ignition and burn rates. Recent research has shown that a Reactivity Controlled Compression Ignition (RCCI) has better scope to provide higher efficiency, lower emissions over a wide range of load and speed. However, it may be noted that HCCI is still only in the development stage and yet to establish itself to replace SI and CI engines.

The thermodynamic cycle of an HCCI engine is close to an Otto cycle. Actually, the concept of HCCI already existed even before electronics came into the picture.

A typical example is the hot-bulb engine. In this, mixing and vaporizing were taken care of in a hot vaporization chamber. The additional heat in the hot chamber along with compression helped combustion to start.

If there is appropriate air-fuel mixture concentration and temperature then the mixture will ignite by itself. The temperature and/or concentration can be changed either by increasing the compression ratio, or pre-heating induction gases. It can also be done by forced induction, or re-inducted exhaust gases. If the ignition is successful then the combustion is very rapid. If the mixture auto ignites too early or with too much chemical energy, combustion will be very fast and there will be a steep pressure rise which can cause damage to the engine. Because of this, HCCI engines are normally operated at lean overall fuel mixtures. As already mentioned, the challenging job in HCCI is the proper control of combustion.

5.5.1 CONTROL

Thus, controlling HCCI combustion is a challenging problem. This is the reason why there is no widespread commercialization. Now comparing conventional SI and CI engines with HCCI engines, it is evident that in both of the conventional engines the timing of combustion can be explicitly controlled. However, in an HCCI engine, the homogeneous mixture of fuel and air is compressed and combustion begins whenever the appropriate conditions are reached. This means that there is no well-defined combustion initiator that could be controlled precisely.

In HCCI engines, to control the combustion, there must be an active control system to precisely monitor the conditions that induce combustion. This may be achieved by employing any one of the following methods

- Varying the compression ratio.
- Varying exhaust gas percentage.
- Varying fuel ignition quality.
- Varying induction temperature.
- Varying valve actuation timing.

These various control methods are briefly discussed in the following sections.

5.5.2 VARYING THE COMPRESSION RATIO

Compression ratio can be classified into two categories: (i) geometric compression ratio and (ii) effective compression ratio. For example,

- The geometric compression ratio, which is the ratio of total volume to clearance volume, can be altered with a movable plunger at the top of the cylinder head.
- The effective compression ratio can be altered by changing the valve timing by incorporating a suitable variable valve actuation mechanism.

Both the approaches mentioned are a little complex and require some additional power to achieve fast responses, and as of today they are quite costly. However, various studies indicate control of HCCI engines by variable compression ratio strategies is more appropriate compared to altering effective compression ratio. Lot of research has gone into studying effects of compression ratio on HCCI combustion.

5.5.3   VARYING EXHAUST GAS PERCENTAGE

In order to vary the exhaust gas percentage there are two possibilities:

1. Retaining hot in-cylinder exhaust gas by altering the valve actuating mechanism.
2. Re-inducting comparatively cool exhaust from the previous cycle through the intake system.

The first one is called hot EGR and the second, cool EGR. Both have been tried. Keeping more exhaust gas in the system has dual effects on HCCI combustion. Hot EGR will increase the temperature of the gases in the cylinder and has a direct effect on the start of ignition whereas cool EGR dilutes the fresh charge, delaying ignition and reducing the chemical energy and engine work. Control of combustion timing in HCCI engines using EGR has also been studied extensively.

5.5.4   VARYING FUEL IGNITION QUALITY

For controlling combustion in HCCI engines, first of all, start of ignition as well as heat release rate should be controlled. This can be achieved by adopting multiple fuels and blending. Examples could be blending of commercial gasoline and diesel or natural gas or ethanol. This can be achieved by the following two ways:

Blending fuels upstream of the engine: Two fuels can be mixed in the liquid phase, one with lower resistance to ignition (such as diesel fuel) and a second with a greater resistance (gasoline). The timing of ignition should be controlled by varying the ratio of these fuels. This fuel mixture is then delivered using either as port or direct injection.

Having two fuel circuits: By adopting a dual fuel concept, fuel A can be injected in the intake duct (port injection) and fuel B can be injected directly into the cylinder. The proportion of these fuels can be used to control ignition, heat release rate and exhaust emissions.

5.5.5   VARYING INDUCTION TEMPERATURE

As mentioned in the previous section, temperature is the major parameter that controls auto-ignition in HCCI engines. Different techniques are adopted to control combustion timing using temperature. The simplest of all the methods uses resistance heaters to vary the inlet temperature. But this approach is not feasible to change the temperature on a cycle-to-cycle basis. Another technique is known as the fast thermal management method, which is achieved by varying the cycle to cycle intake charge temperature by rapidly mixing hot and cold air streams. It is found to be very costly to implement and has limited range of application.

5.5.6   VARYING VALVE ACTUATION

Varying valve actuation (VVA) is found to provide an extended range of application for HCCI engines by giving finer control over the temperature-pressure-time history within the combustion chamber. VVA can be achieved by the following two methods:

> *Controlling the effective compression ratio*: A VVA system on intake can control the point at which the intake valve closes. If this is retarded past BDC, then the compression ratio will change, altering the in-cylinder pressure-time history prior to combustion.
>
> *Controlling the amount of hot exhaust gas retained in the combustion chamber*: A VVA system can be used to control the amount of hot internal exhaust gas recirculation (EGR) within the combustion chamber. This can be achieved either by including valve re-opening and changes in valve overlap. By balancing the percentage of cool EGR with the hot internal EGR generated by a VVA system, it may be possible to control the in-cylinder temperature.

Electro-hydraulic and camless VVA systems are more attractive to provide a great deal of control over the valve event. However, manufacturing of components for such systems is currently complicated and expensive. Mechanical systems with variable lift and duration, however, are far cheaper and less complicated. If the desired VVA characteristic is known, then it is relatively simple to configure such systems to achieve the necessary control over the valve lift curve.

5.5.7   POWER

In conventional engines power is controlled by the amount of fuel and they can withstand higher power because of slow heat release rate. However, in HCCI engines the entire mixture burns nearly simultaneously and increasing the fuel-air ratio will result in even higher peak pressures and heat release rates. In addition, many of the viable control strategies for HCCI require thermal pre-heating of the charge which will reduce the density and hence the mass of the fuel-air charge in the combustion chamber, reducing power. These factors make it a challenging task to improve the power output in HCCI engines.

There are three possibilities to increase the power of HCCI engines:

1. To use fuels with different auto-ignition properties. This will lower the heat release rate and peak pressures and will make it possible to increase the equivalence ratio.

2. To thermally stratify the charge. This will make different points in the compressed charge have different temperatures and burn at different times, lowering the heat release rate.
3. To run the engine in HCCI mode, only at part load conditions and run it as a diesel or Spark Ignition engine, at full or near full load conditions.

Since many advantages are there in thermal stratification in the compressed charge, the last approach is being pursued more vigorously.

5.5.8 EMISSIONS

Compared to conventional gasoline and diesel engines, HCCI engines operate on lean mixtures and therefore, the peak temperatures are lower. The lower peak temperatures reduce the formation and emission of NO_x which is very much lower. However, the low peak temperatures also lead to incomplete burning of fuel, especially near the walls of the combustion chamber. This can lead to high carbon monoxide and hydrocarbon emissions. An oxidizing catalyst is required to remove the regulated species because the exhaust is still oxygen rich.

5.5.9 DIFFERENCE IN ENGINE KNOCK

In conventional spark ignition engines, knock or pinging occurs when some of the unburnt gases ahead of the flame spontaneously auto-ignite. This causes a shock wave to traverse from the end gas region and an expansion wave to traverse into the end gas region. The two waves reflect off the boundaries of the combustion chamber and interact to produce high amplitude standing waves.

In HCCI engines, ignition occurs due to piston compression and the entire reactant mixture ignites (nearly) simultaneously. Since there are very little or no pressure differences between the different regions of the gas, there is no shock wave propagation and hence, no knocking. However, at high loads (i.e. high fuel-air ratios), knocking is a remote possibility even in HCCI.

5.5.10 ADVANTAGES AND DISADVANTAGES OF HCCI ENGINE

Advantages

- They include considerable fuel savings.
- Current emissions norms can be met.
- They can operate at diesel-like compression ratios.
- Efficiencies are much higher than conventional SI engines.
- Due to homogeneous mixing cleaner combustion and lower emissions are achieved.
- Peak temperatures are lower than the typical SI engines.
- NO_x and soot emissions are almost negligible.

- They can operate on both gasoline and diesel, and almost all alternative fuels.
- Throttling losses are less.

Disadvantages

- High in-cylinder peak pressures may cause damage to the engine.
- Fast heat release and pressure rise rates contribute to engine wear.
- The auto-ignitions are difficult to control, unlike the SI and CI engine.
- HCCI engines have a small power range.
- Constraints at low loads are due to lean flammability limits.
- Problem at high loads are due to in-cylinder pressure restrictions.
- CO emissions are higher compared to conventional engines.
- Pre-catalyst hydrocarbon emissions are higher.

5.6   STRATIFIED CHARGE ENGINE

Gasoline and diesel engines are the two well-established types of internal combustion engines. However, they have their own limitations such as Otto engines having poor part load efficiency, but they provide very good full load power characteristics. Similarly, diesel engines, having higher weight-to-power ratio and low smoke-limited power, give good part load characteristics. Higher losses and problems of maintenance in diesel operation are caused by use of higher compression ratios. Both full load power and part load efficiency are very important for an automotive engine during its operating life. It should be noted that they work most of the time, under part load conditions. Speed, acceleration and other characteristics determine the maximum power requirement of the vehicle. Hence, an engine has to be developed with the advantages of both the gasoline and diesel engines with avoiding their disadvantages. An engine that is midway between a heterogeneous charge Compression Ignition engine and homogeneous charge Spark Ignition engine is being attempted and is known as a *stratified charge engine*.

Different mixture strengths of air and fuel such as relatively rich mixture at the vicinity of the spark plug and a leaner mixture in the rest of the combustion chamber is provided at various places in the combustion chamber for charge stratification. But the overall mixture is leaner. Thus, the stratified charge engine can be said to be a Spark Ignition internal combustion engine where the mixture in the zone of spark plug is very much richer than that of the rest of the combustion chamber.

5.6.1   ADVANTAGES OF BURNING LEANER OVERALL FUEL-AIR MIXTURES

Higher thermodynamic efficiency: The SI engine output is controlled by a throttle where the mixture strength is kept nearly constant and the quantity is varied during the induction of suction stroke. The output of the unthrottled diesel engine is controlled in every cycle by varying the amount of fuel injected into a constant amount of air. Thus, the gasoline engine operates within a very narrow range of fuel-air ratios and the diesel engine operates over a much wider range of lean mixture strength.

The thermodynamic efficiency of the Otto cycle is given by

$$\eta = 1 - \frac{1}{r^{\gamma-1}}$$

The value of γ for air is 1.4, the chemically correct carburetted fuel-air mixture is around 1.3 and a leaner mixture is between 1.3 and 1.4. From this equation, it is clear that lean mixture would result in slightly higher thermodynamic efficiency; refer to curve 1 in Figure 5.4. The range of mixture strength is quite narrow (i.e. between 0.8 and 1.2) for SI engines, and the range is quite wide (i.e. between 0.1 and 0.8) for CI engines; refer to curve 2 in Figure 5.4. This is also a reason for better efficiency at part load in diesel engines. The unthrottled diesel engine has an excess air of about 20 to 40% at full load and progressively increases at part load as reduced amounts of fuel are injected. This is not so with gasoline engines where the mixture strength is almost constant.

The variation of theoretical efficiencies for different mixture strengths under ideal conditions of Otto, diesel and stratified charge engines (curve 3) is shown in Figure 5.4. It can be seen that at the comparable compression ratio of gasoline, the part load efficiency of the stratified charged engine is much better and almost reaches the

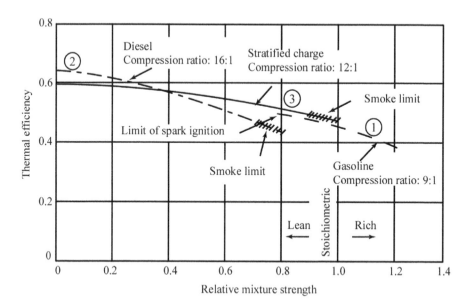

FIGURE 5.4 Theoretical efficiencies of Otto, diesel and stratified charge engine cycles for various mixture strengths.

Source: Reproduced from the book *Internal Combustion Engine* by V. Ganesan, fourth edition, with permission from McGraw Hill Education New Delhi.

efficiency of the diesel engine. However, the higher compression ratio is the advantage for the diesel engines. Hence, the performance level approaching that of a CI engine can be achieved with a slightly higher compression ratio and leaner mixtures in an SI engine.

This will help to overcome the disadvantages faced in diesel operations, such as good combustion over a wide range of mixture strength, high compression ratios for better starting and poor air utilization. However, a higher weight-to-power ratio, higher mechanical losses and greater maintenance are the problems faced with the use of a high compression ratio. The poor air utilization results in smoky operation at higher loads and poor fuel economy.

Another important point to note from Figure 5.4 is the limited range of mixture strength of 0.8 to 1.2 that can be used in a gasoline engine. Propagation of flame throughout the mixture is a must for complete combustion. Hence, only those mixtures through which flame propagation is possible can be used. Mixture strength below a relative fuel-air ratio of 0.8 for a single-cylinder engine will result in misfiring. Further, in a multi-cylinder engine this value is 0.85, due to imperfect distribution among the cylinders. The maximum output, mixture strength is about 1.05 to 1.1 and 1.2 for a multi-cylinder engine. At a relative fuel-air ratio of 0.85 to 0.9, minimum specific fuel consumption is obtained and flame propagation becomes slower and fluctuations occur from cycle to cycle, which limits the use of leaner mixtures. A relative fuel-air ratio as low as 0.2 (corresponding to idling conditions) can be used in stratified charge engines.

Another disadvantage of the gasoline engine is at nearly constant mixture strength operation, it results in almost constant peak cycle temperature over the full load range. This implies that the losses due to dissociation at high temperature, heat transfer and variable specific heats would be much higher at part loads. These shortcomings of the gasoline engine can be overcome by using charge stratification.

Reduced air pollution: Flame quenching at the combustion chamber walls in SI engines produces higher hydrocarbons in exhaust. This quenching effect is drastically reduced in a stratified charge engine due to its use of overall lean mixture and almost pure air will be present near the combustion chamber walls at part loads, thereby reducing amounts of NO_x and CO and lowering HCs. Reduced residence time under high temperature and pressure conditions causes less knocking in stratified charge engines. The following are the distinct advantages of using charge stratification: (i) no throttling losses, (ii) multi-fuel capability and (iii) lower knocking tendency.

5.6.2   Methods of Charge Stratification

The stratified charge engines can be categorized according to the method of formation of the heterogeneous mixture in the combustion chamber as those using:

1. Fuel injection and positive ignition (including swirl stratified charge engines).
2. Carburetion alone.

5.6.3 STRATIFICATION BY FUEL INJECTION AND POSITIVE IGNITION

Ricardo was the first to attempt in 1922 to obtain charge stratification. Figure 5.5 gives the details. An auxiliary spray forms a relatively rich mixture at the vicinity of the spark plug and another spray of fuel forms a leaner mixture in the combustion chamber.

This arrangement could give efficiencies as high as 35% due to the overall very lean operation with a wide range of mixture strengths. However, the engine did not work satisfactorily at loads higher than 50% of full load. This may be due to very rich mixture near the spark plug.

Prechamber stratified charge engine: A small prechamber fitted with an injector and a spark plug as shown in Figure 5.6 was later used by Ricardo.

In this arrangement, the injector supplies fuel which forms a rich mixture near the spark plug and a carburettor supplies lean mixture to the main combustion chamber. The auxiliary charge burns in the prechamber and passes through its throat to the main chamber and thus burns the lean mixture present. However, this method involves the following problems during engine operations:

1. Getting good performance at the full load operation.
2. Incomplete scavenging of the prechamber.
3. Improper burning of rich mixture at full load due to improper fuel distribution.
4. Loss of thermal efficiency due to throttling.

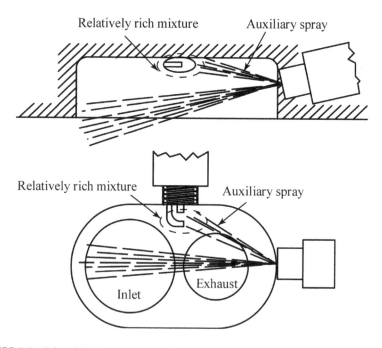

FIGURE 5.5 Ricardo's first charge stratification approach.

Source: Reproduced from the book *Internal Combustion Engine* by V. Ganesan, fourth edition, with permission from McGraw Hill Education New Delhi.

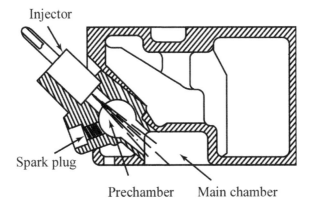

FIGURE 5.6   Ricardo's prechamber stratification.

Source: Reproduced from the book *Internal Combustion Engine* by V. Ganesan, fourth edition, with permission from McGraw Hill Education New Delhi.

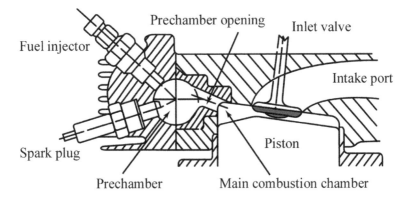

FIGURE 5.7   Volkswagen PCI stratified charge engine.

Source: Reproduced from the book *Internal Combustion Engine* by V. Ganesan, fourth edition, with permission from McGraw Hill Education New Delhi.

5.6.4   VOLKSWAGEN PCI STRATIFIED CHARGE ENGINE

A Volkswagen PCI stratified charge engine consists of a spherical unscavenged pre-chamber, which comprises approximately 25–30% of the compression volume. It is linked through a flow passage by a relatively large prechamber opening to the disc-shaped main combustion chamber. A slight swirl is induced by the intake port as it does not have squish surface. The injection nozzle and spark plug are arranged in sequence in the flow direction during compression for the spark plug to receive a mixture of incoming air with fuel already dispersed in air and to avoid over-enrich-ment at the spark plug. This arrangement is illustrated in Figure 5.7.

The total fuel volume is divided and is injected partly into the prechamber and partly into the intake manifold. However, a relatively rich mixture is made available under all operating conditions at the spark plug and mixture strength is adjusted and introduced into the main combustion chamber to achieve load regulation.

The main advantage of this system is that the fuel injection timing needs not to be varied as compared to other stratification processes. At higher speeds, the octane requirement is comparatively low and from an exhaust emission point of view, it can be easily operated on very lean mixtures (i.e. relative air-fuel ratios greater than 2).

5.6.5 BRODERSON METHOD OF STRATIFICATION

Performance at part load and full load are not the same in gasoline engines. These can be tackled either by using fuel injection alone and controlling the mixture strengths in auxiliary and main combustion chambers by means of injection timings or by using carburetion alone to obtain stratification with or without the prechamber.

The Broderson method of charge stratification consists of an engine with a divided combustion chamber. The small or auxiliary chamber contains the injector, the intake valve and the spark plug. This is illustrated in Figure 5.8.

Entire fuel is injected into the auxiliary chamber. The fuel can be injected after bottom dead centre (bdc) (i.e. after the start of compression stroke at light loads). During this time, the air moves from the main chamber into the auxiliary chamber and the fuel injected into the auxiliary chamber will remain there and will form an ignitable mixture near the spark plug. The main chamber will contain only air or very lean mixture. This lean mixture in the main chamber gets burnt due to the torch effect by the blast of flame from the auxiliary chamber. The main chamber must be so designed to ensure proper turbulence for mixing of the hot burning mixture coming out of the auxiliary chamber with the mixture and also for rapid and complete burning of this lean mixture. At full load the fuel is injected before bdc (i.e. during the suction stroke). It gets thoroughly mixed with the entire air to form a mixture of uniform strength throughout the combustion space. The flame produced by the spark plug will spread into both the chambers as in the normal spark ignition operation.

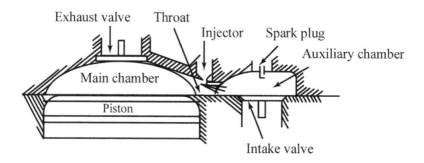

FIGURE 5.8 Broderson method of stratification.

The advantages of the Broderson method are the avoidance of the pumping losses due to throttling, knockless operation at full load and higher part load efficiency due to lean mixture. Emissions will also be less as a very lean mixture remains in the main chamber at part load operations. Compared to conventional gasoline engines; this method gives lower exhaust temperatures. This type of divided chamber stratified charge engine can burn almost any fuel such as diesel, gasoline, propane or kerosene fuels. It has multi-fuel capability inherent in it and thus little sensitivity to octane number.

The disadvantages of this method are:

1. Great care is required in the design of the auxiliary chamber and main chamber as it requires greater attention for matching the injection and ignition timing.
2. A hollow and sharp sound is produced at idling due to rapid expulsion of gases from the auxiliary chamber and this noise increases with increase in load.

5.6.6 CHARGE STRATIFICATION BY SWIRL

It is known that a wide range of air-fuel ratio can be burnt by properly tuning the injection system in an open combustion chamber itself. Thereby, the disadvantages in a divided chamber engine can be avoided. To obtain this, the fuel injection and air swirl are to be properly matched in an open combustion chamber to give charge stratification. Over the years, quite a few designs of stratified charge engines utilizing air swirl in an open combustion chamber have been developed. Some of them are

1. Ford combustion process (FCP).
2. Ford PROCO.
3. Texaco combustion process (TCP).
4. Witzky swirl stratification process.

5.6.7 FORD COMBUSTION PROCESS (FCP)

Ford combustion process attempts to obtain part load stratification by proper coordination of fuel injection and air swirl ensuring positive ignition and also maintains the full load potential of the homogenous fuel-air mixture. However, spray characteristics play an important role.

Layout of the Ford combustion process is given in Figure 5.9. The injector is located radially at a specified angle (56°) as shown in the Figure 5.9a. This location is found to provide good mixing at all loads and speeds.

A long electrode used in the spark plug is so positioned that the spark gap is about 10 mm ahead of the injector along its centre line. A recessed and shrouded intake valve Figure 5.9b is used to impart the appropriate air motion with high turbulence and the stationary shroud reduces the anti-swirl in air flow. A directional intake port is used to properly guide the intake flow. Therefore, intake port design is very critical for this system. The working process is as follows.

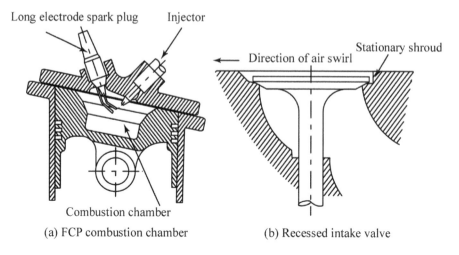

(a) FCP combustion chamber   (b) Recessed intake valve

FIGURE 5.9   Ford combustion process.

Source: Reproduced from the book *Internal Combustion Engine* by V. Ganesan, fourth edition, with permission from McGraw Hill Education New Delhi.

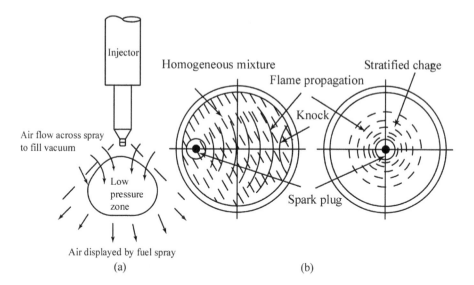

FIGURE 5.10   Air flow pattern in FCP combustion chamber.

Source: Reproduced from the book *Internal Combustion Engine* by V. Ganesan, fourth edition, with permission from McGraw Hill Education New Delhi.

As shown in Figure 5.10a, a low-pressure region in front of the injector is created as the injection of the fuel spray moves a substantial amount of air forward. Air behind the injector tip moves into this low-pressure region and helps to drag some of the smaller droplets. They get vaporized and form an ignitable air-fuel mixture some 5 to 10 mm ahead of the injector tip. Heavy droplets do not get closer to the spark plug due to higher inertia to drag forces. This ignitable mixture, which is at or near the bore centre, rotates without drifting away from the spark plug and also it gets mixed properly by the swirling air.

Possible stratification in the combustion chamber at different loads and speeds is shown in Figure 5.10b. It can be seen that there is a significant difference in the mixture distribution pattern at part load and full load. Therefore, injection and ignition timings should be changed as per the load and speed for proper and complete combustion. The injection timing is advanced with increase in load for better air utilization. The spark timing is adjusted to achieve maximum cycle efficiency. This requires a spark advance mechanism as in normal gasoline engine. Though flame velocity increases with engine speed it is not good enough to give appropriate heat release.

5.6.8 FORD **PROCO**

The PROCO engine is the second generation of Ford combustion process (FCP). This was developed to give fuel economy and maximum power close to a carburetted engine. The PROCO process aims at minimizing the exhaust emissions with maximum fuel economy. Unlike FCP, it uses air throttling for this purpose.

Intake port is designed in a shape to impart a swirl around the cylinder bore axis. This is illustrated in Figure 5.11. The swirl speed is about 4 to 5 times the crankshaft speed in the engine combustion chamber situated concentrically in the piston with about 70% squish area.

The combustion chamber is cup-shaped with about 70% squish area and situated concentrically in the piston and the injector is located close to the centre of the

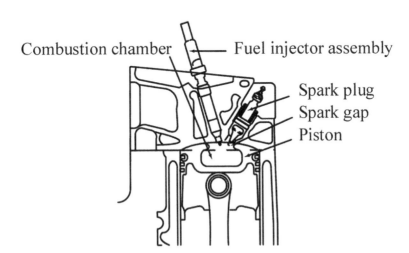

FIGURE 5.11   Ford PROCO engine.

Source: Reproduced from the book *Internal Combustion Engine* by V. Ganesan, fourth edition, with permission from McGraw Hill Education New Delhi.

cylinder bore. The swirling charge is compressed into the combustion chamber. The compression ratio is 11:1. Fuel is injected as a soft, wide-angle conical spray and is low penetrating during the compression stroke. Hence, the fuel is a rich mixture at the centre and surrounded by a lean mixture and excess air. The spark plug is located either near the bore centreline or just above the spray. It ignites the charge near top dead centre. Combustion takes place faster in rich mixture and then spreads gradually into the leaner regions.

5.6.9   Texaco Combustion Process (TCP)

Figure 5.12 shows the Texaco combustion process. In this process, the inlet manifold is suitably designed for a high degree of swirl which is imparted to the incoming air. The fuel is injected towards the end of the compression stroke and in downstream directions across the swirling air so that the entire air is impregnated by fuel.

Normally fuel is injected about 30° before tdc or just after the initiation of the spark. A glow plug or a spark plug for ignition is positioned close to the injector and directly downstream, within the swirling air and outside the spray envelope. The first part of the injected fuel mixes with air during its passage to the spark plug and forms a flame front and then the remaining fuel is continuously injected after mixing with the swirling air into this already established flame front. There must be a good coordination between air swirl and injection with guaranteed ignition for proper operation of TCP.

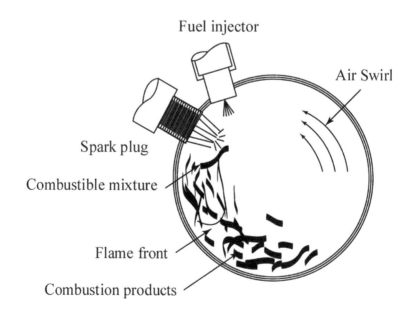

FIGURE 5.12   Texaco combustion process engine.

Source: Reproduced from the book *Internal Combustion Engine* by V. Ganesan, fourth edition, with permission from McGraw Hill Education New Delhi.

There are two important characteristics of this process.

1. Almost knock-free operation is due to the quite small residence time of the fuel-air mixture in the combustion zone. It gives the engine the multi-fuel capability due to its insensitivity to the octane numbers of the fuel.
2. Very lean mixtures can be burnt easily by terminating the fuel injection before the complete utilization of the swirling air.

At part loads, a small patch of ignitable mixture is near the spark plug whereas the rest of the chamber has very lean mixture. As the load increases, the area of this patch also increases and becomes just like the homogeneously charged gasoline engine at full loads.

The advantages of TCP are that a wide range of fuels ranging from premium gasoline to high cetane diesel fuel can be used. It gives good performance over the whole range of speed and load. Due to its cup-shaped combustion chamber, the part load efficiency is better and the starting and warm-up characteristics of TCP are also quite good. The inherent knock resistance of TCP allows the use of higher compression ratio or turbocharging. The lean overall mixtures can be used. The exhaust smoke is also very low.

Timing of the fuel injection is the major difference between the TCP and Ford PROCO systems. Combustion occurs when fuel is injected into the compressed air and is immediately ignited by the spark plug near the end of the compression stroke in a Texaco engine whereas the fuel is injected into the cylinder over a longer period of time in a Ford engine,

5.6.10 WITZKY SWIRL STRATIFICATION PROCESS

The Witzky swirl stratification process is basically an unthrottled spark ignition process with fuel injection with no carburetion. This is illustrated in Figure 5.13. The

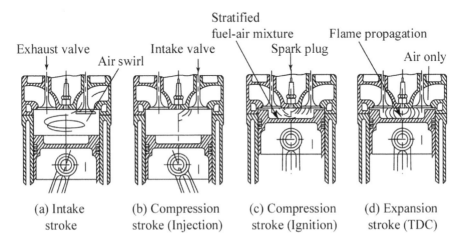

FIGURE 5.13 Witzky swirl stratification process.

Source: Reproduced from the book *Internal Combustion Engine* by V. Ganesan, fourth edition, with permission from McGraw Hill Education New Delhi.

intake port is appropriately designed for high velocity swirl to impart the intake air during the suction stroke. The fuel is injected at some suitable angle against the swirl direction during the compression stroke. The swirling air forces the fuel droplets to follow a spiral path by virtue of drag forces and directs them towards the centre of the combustion chamber. A spark plug is used for initiating the combustion. This produces a good degree of stratification—a rich ignitable mixture near the spark plug over the full load range and leaner mixture away from the spark plug. Close to the wall almost pure air is present. The thickness of this pure air layer decreases as the load is increased.

5.6.11 HONDA CVCC ENGINE

Even though stratifications are being attempted by injection, there are attempts to achieve stratification by carburetion alone. Honda compound vortex-controlled combustion (CVCC) is a bold attempt in this direction. The Honda CVCC engine uses a three-barrel carburettor for charge stratification. The engine has two chambers, a prechamber and a main chamber. The prechamber is filled with a rich mixture and is connected to the main chamber that is filled with lean mixture. Chambers are filled using carburettors. Figure 5.14 shows the Honda CVCC engine. The basic structure of the engine is similar to a conventional four-stroke gasoline engine. But it has an auxiliary combustion chamber around the spark plug and also a small additional intake valve fitted to each cylinder. Two separate intake valves are used on each cylinder. One valve is located in the prechamber and the other in the main chamber. The smallest venturi of the three-barrel carburettor supplies a rich mixture to each prechamber. The other two ventures supply a very lean mixture to the main chamber.

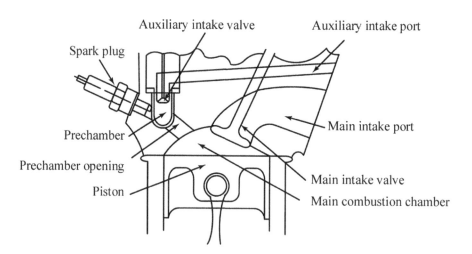

FIGURE 5.14   Honda CVCC engine combustion chamber.

Source: Reproduced from the book *Internal Combustion Engine* by V. Ganesan, fourth edition, with permission from McGraw Hill Education New Delhi.

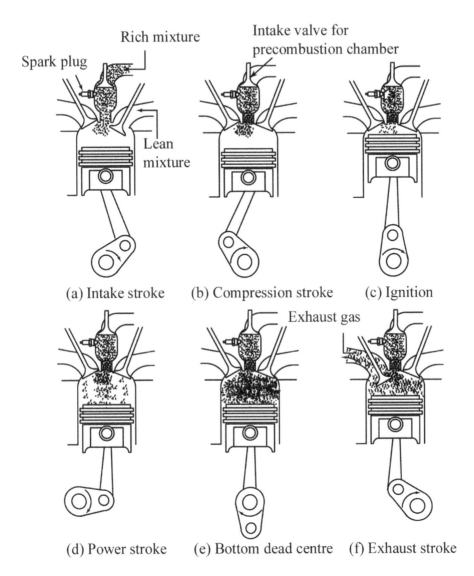

FIGURE 5.15   Combustion sequence of Honda CVCC engine.

Source: Reproduced from the book *Internal Combustion Engine* by V. Ganesan, fourth edition, with permission from McGraw Hill Education New Delhi.

A conventional ignition system and spark plugs (one plug per prechamber) are used to initiate combustion. Figure 5.15 shows the schematic of the combustion system. The combustion sequence is as follows:

- A large amount of lean mixture is supplied through the main intake valve and a small amount of rich mixture through auxiliary intake valves. The overall mixture is lean in the system.

- There is a rich mixture around the spark plug, a moderate mixture in the vicinity of the outlet of the auxiliary chamber and a lean mixture in the rest of the mean combustion chamber at the end of the compression stroke.
- The rich mixture in the auxiliary chamber is ignited by the spark plug.
- The moderate mixture at the outlet of auxiliary chamber is ignited which in turn ignites the lean mixture in the main combustion chamber.
- The lean mixture continues to burn slowly.
- The temperature of the burnt gas remains relatively high for a long duration.

The overall air-fuel ratio of the CVCC engine is much leaner than the stoichiometric ratio. It has good emission characteristics. Low peak combustion temperatures at lean mixtures and sufficiently high temperature during exhaust results in reduced NO_x, CO and HC.

5.6.12 ADVANTAGES AND DISADVANTAGES OF STRATIFIED CHARGE ENGINES

Advantages

- It can use a wide range of fuels.
- It has low exhaust emission levels.

Disadvantages

- Charge stratification results in reduced power for a given engine size.
- Its weight is higher for the comparable power compared to conventional engines.
- Its manufacturing cost is higher.
- Its durability and reliability are not well established.

ACKNOWLEDGEMENT

The authors are thankful to McGraw Hill Education for giving permission to reproduce figures from the book *Internal Combustion Engine* by V. Ganesan, fourth edition.

BIBLIOGRAPHY

1. Arthur Pound, The Turning Wheel; The Story of General Motors Through Twenty-Five Years, 1908-1933, Garden City, N.Y., Doubleday, Doran & Company, Inc., 1934.
2. Alkidas A. C. Combustion advancements in gasoline engines, *Energy Conversion and Management*, 48 (2007), pp. 2751–2761.
3. Alkidas A. C. Tahry, Contributors to the fuel economy advantage of DISI engines over PFI engines, SAE 2003–01–3101, 2000.
4. Bartolini C. M., Vicenzi G., Chiatt G. Performance analysis of two stroke engine with direct injection, SAE 931508, 1993.
5. Blair P. Design and simulation of two stroke engines, SAE 1996 ISBN, 1-56091-685-0, 1996.

6. Block B., Westphal H., Oppermann W., Hentschel W., Henning H., Kutschera I. Optical detection of the combustion produced by the pre-injected fuel in a DI diesel engine, 2002, SAE Paper 2002–01–2667.

7. Carson C. E., Kee R. G., Kenny R. G., Lehaman S., Zwahir S. Ram Tunned and air assisted fuel injection system applied to two stroke engine, 1995, SAE 950269.

8. Celik M. B., Ozdalyan B. Gasoline direct injection. In: Siano D., editor. *Fuel Injection*, Sciyo; 2010, www.intechopen.com, *Sadhana* Vol. 40, Part 6, September 2015, pp. 1937–1954. c Indian Academy of Sciences.

9. Cheng W. T., Sub H. J., Xie X., Chia L. M., Henein N. A., Schwarz E., Bryzik W. Direct visualization of high pressure diesel spray and engine combustion, 1999, SAE Paper 1999–01–3496.

10. Costa M., Sorge U., Allocca L. Increasing energy efficiency of a gasoline direct injection engine through optimal synchronization of single or double injection strategies, *Energy Conversion and Management* (2012).

11. Costa M., U. Sorge, Allocca L. CFD optimization for GDI spray model tuning and enhancement of engine performance, *Advances in Engineering Software*, 49 (2012), pp. 43–53.

12. Desantes J. M., Arregle J., Lopez J. J. and Garcia A. A comprehensive study of diesel combustion and emissions with post-injection, 2007, SAE Paper 2007–01–0915.

13. Drake M. C., Haworth D. C. Advanced gasoline engine development using optical diagnostics and numerical modelling, *Proc of the Combustion Institute*, 31 (2007), pp. 99–124.

14. Drake M. C., Fansler T., Lippert A. M. Stratified-charge combustion: Modelling and imaging of a spray-guided direct-injection spark-ignition engine, *Proceedings of the Combustion Institute*, 30 (2005), pp. 2683–2691.

15. Drake M. C., Hawotrh D. C. Advanced gasoline engine development using optical diagnostics and numerical modelling, *Proceedings of the Combustion Institute*, 31 (2007), pp. 99–124.

16. Eblen E. and Stumpp G. Beitrag des Einspritz systems zur Verbesserung des Diesel motors, BoschTechn. Berichte Band 6, Heft 2Endres H, Dumholz M and Frisse P 1994 Pre-injection: A measure to optimize the emission behavior of DI-diesel engines, 1978, SAE Paper 940674.

17. Rottenkolber G., Gindele J., Raposo J., Dullenkopf K., Hentschel W., Spicher U. Spray analysis of a gasoline direct injector by means of two-phase PIV, *Experiments in Fluids*, 32 (2002) 710–721 Springer-Verlag 2002.

18. Ganesan V. *Internal Combustion Engines*, Tata McGraw-Hill Education Pvt. Ltd, 2013, New Delhi.

19. Gupta A., Jonathan S., Ramzy G., Abdel-Gayed. Downsizing of a naturally aspirated engine to turbocharged gasoline direct injection variable valve timing engine, *International Journal of Engineering and Applied Sciences*, 4 (6) (Dec 2013).

20. Heimberg W. Fitch pressure surge injection system, 1993, SAE paper 931508.

21. Henein N. A., Lai M. C., Singh I., Wang D., Liu L. Emissions trade off and combustion characteristics of a high speed direct injection diesel engine, 2001, SAE Paper 2001–01–0197.

22. Hentschel W. Optical diagnostic for combustion process development of direct injection engines, *Proceedings of the Combustion Institute*, 28 (2000), pp. 1119–1135.

23. Ishiwata H., Ohishi T., Ryuzaki K., Unoki K., Kitahara N. A feasibility study of pilot injection in TICS, 1994, SAE Paper 940195.

24. Kohketsu S., Mori K., Kato T., Sakai K. Technology for low emission, combustion noise and fuel consumption on diesel engine, 1994, SAE Paper 940672.

25. Kong S. C., Karra P. K. Diesel emission characteristics using high injection pressure with converging nozzles in a medium-duty engine, 2008, SAE Paper 2008–01–1085.

26. Koyanagi K., Oing H., Renner G., Maly R. Optimizing common rail-injection by optical diagnostics in a transparent production type diesel engine, 1999, SAE Paper 1999–01–3646.

27. Kumerave K., Prem Ananad B., Anushka K., Saravanan C. G. Effect of air fuel ratio and injection timing on gasoline direct injection engine performance. 23rd national conference on IC engine and combustion (NCICEC2013) SVNIT, Surat, India 13–16 Dec 2013.

28. Miyaki M., Fujisawa H., Masuda A., Yamamota Y. Development of new electronically controlled fuel injection system ECD-U2 for diesel engines, 1991, SAE Paper 910251.

29. Nakakita K., Kondoh T., Ohsawa K., Takahashi T., Watanabe S. Optimization of pilot injection pattern and its effect on diesel combustion with high-pressure injection, *JSME Int. J. Series B—Fluids Thermal Eng.*, 37 (1992), pp. 966–973.

30. Nakashimaa T., Basaki M., Saitob K., Furuno S. New concept of a direct injection SI gasoline engine: A study of stratified charge combustion characteristics by radical luminescence measurement, *JSAE Review*, 24 (2003), pp. 17–23.

31. Nohira H. Development of Toyota's direct injection gasoline engine, AVL Engine and Environment Conference, 1997, pp. 239–249.

32. Park C. W., Oh C., Kim S. D., Kim H. S., Lee S. Y., Bae C. S. Evaluation and visualization of stratified ultra lean combustion characteristics in a spray guided type gasoline direct injection engine, *J of Automotive Technology*, 15 (4) (2014), pp. 525–533

33. Park C., Kim S., Kim H., Moriyoshi Y. Stratified lean combustion characteristics of a spray-guided combustion system in a gasoline direct injection engine, *Energy*, 41 (2012), pp. 401–407.

34. Thirouard M., Mendez S., Pacaud P., Chmielarczyk V., Didier A., Chrsitophe G., Frédéric L., Bertrand B., et al. Potential to improve specific power using very high injection pressure in HSDI Diesel engines, 2009, SAE Paper 2009–01–1524.

35. Riaud J. C., Lavoisier F. Optimizing the multiple injection settings on an HSDI diesel engine. In: THIESEL conference on Thermo and Fluid Dynamic Processes in Diesel Engines, 2002.

36. Rinolf R., Imarisio R., Buratti R. The potential of a new common rail diesel fuel injection system for the next generation of DI diesel engines, 16 Internationales Wiener Motoren symposium, VDI-Verlag Reihe 12 Nr. 239, 1995.

37. Bosch R. 2006 *GmbH, Diesel-Engine Management*. John Wiley and Sons Inc., New Jersey USA.

38. Rotondi R., Bella G. Gasoline direct injection spray simulation, *Int, J Thermal Sci*, 45 (2006), pp. 168–179.

39. Salber W., Wolters P., Esch T., Geiger J., Dilthey J. Synergies of variable valve actuation and direct injection, 2002, SAE 2002–01–0706.

40. Schubiger R., Bertola A., Boulouchos K. Influence of EGR on combustion and exhaust emissions of heavy duty DI-diesel engines equipped with common-rail injection systems, 2001, SAE Paper 2001–01–3497.

41. Shimada T., Shoji T., Takeda Y. The effect of fuel injection pressure on diesel engine performance, 1989, SAE Paper 891919.

42. Shundoh S., Komori M., Tsujimura K., Kobayashi S. NOx reduction from diesel combustion using pilot injection with high pressure fuel injection, 1992, SAE Paper 920461.

43. Tow T., Pierpont D. A., Reitz R. D. Reducing particulates and NOx emissions by using multiple injections in a heavy duty DI diesel engine, 1994, SAE Paper 940897.

44. Tsurushima T., Zhang L., Ishi Y. A study of unburnt hydrocarbon emissions in small DI diesel engine, 1999, SAE Paper 1999–01–0512.
45. Uchida N., Shimokawa K., Kudo Y., Smimoda M. Combustion optimization by means of common rail injection system for heavy duty diesel engines, 1998, SAE Paper 982679.
46. Zhang L. A study of pilot injection in a DI diesel engine, 1999, SAE Paper 1999–01–3493 nine microns as opposed to 27 microns in a high pressure direct injection system.
47. Zhao F., Lai M. C., Harrington D. L. Automotive spark-ignited direct-injection gasoline engines, *Progress in Energy and Combustion Science*, 25 (1999), pp. 437–562.

6 Design of Downdraft Biomass Gasifier Systems and Naturally Aspirated Producer Gas Burners

M.R. Ravi and Sangeeta Kohli

CONTENTS

6.1 INTRODUCTION

Gasification is a thermochemical process by which a solid fuel is partially oxidised at elevated temperatures (600–1000° C) in a limited supply of oxidiser (15–30%), to yield a combustible gaseous fuel whose heating value ranges from 3–6 MJ/Nm^3 if the oxidiser is air and 10–15 MJ/Nm^3 if it is pure oxygen, with biomass as feedstock [1]. Reactors in which this is accomplished are known as *gasifiers*. Depending on

DOI: 10.1201/9781003049005-6

the flow configuration of air and feedstock, gasifiers are classified as Moving Bed Gasifiers (ironically, also known as fixed bed gasifiers), Fluidised Bed Gasifiers or Entrained Flow Gasifiers. The category of our interest in this chapter is the moving bed gasifiers, which in turn are classified as updraft, downdraft and cross-draft gasifiers, based on the direction of gas flow with reference to the biomass flow, which is always downwards in this category of gasifiers due to gravity. In updraft gasifiers, air is supplied from below the fuel bed, and after gasification in the bed, the producer gas leaves from the top of the gasifier, thus making it a counterflow configuration. In cross-draft gasifiers, air is supplied from one side of the fuel bed, and the producer gas leaves from the opposite side of the bed, making it a cross-flow configuration. The configuration of interest to this chapter is the downdraft gasifiers, where air is supplied through nozzles (also known as *tuyeres*) in the mid-height region of the gasifier. The gas escapes from the bottom of the gasifier, making it a co-flow configuration [2], [3].

The research group under Prof. Mukunda at the Indian Institute of Science has been the leading contributor in India, both to the design of gasification systems and to the scientific literature on gasifiers (Dasappa et al. [4]; Mukunda et al. [5], [6]). There is only a limited amount of literature, from very few research groups, on the design aspects of gasifiers.

Producer gas generated by these gasifiers can be used for shaft power generation or for thermal applications. For the latter, design of suitable burners that can ensure efficient and clean combustion of producer gas is essential. Aerated burners for gaseous fuels are common both in industry and in domestic use. These are classified as fully aerated, partially aerated and non-aerated burners, based on what fraction of the combustion air is provided upstream of the combustion region in the burner. Jones [7] presents methods to design domestic burners for typical hydrocarbon gaseous fuels in detail. Designs of large burners for use with gasifiers are presented in a few papers [8], [9]. However, no details of the design procedure for producer gas burners is available.

This chapter reviews the published work pertaining to the design of downdraft biomass gasification systems and burners. The write-up draws substantially from the experience of the authors' research group in designing gasification systems [10], [11]; and producer gas burners [12].

Section 6.2 introduces downdraft biomass gasifiers and design parameters used for their characterisation. Section 6.3 deals with physical and chemical phenomena of gasification with reference to a downdraft biomass gasifier. Section 6.4 outlines the requirements from a good gasification reactor in view of the physical and chemical phenomena. Section 6.5 discusses the requirements of fuel properties and preparation for good performance of a gasification reactor. Section 6.6 discusses the design process for a downdraft biomass gasification system, based on available literature and the experience of the author's research group.

Section 6.7 discusses the design procedures available in the literature for gas burners, where the driving force for entrainment of air is predominantly the kinetic energy of the gas stream. Section 6.8 presents the modifications to this procedure in order to incorporate the effect of buoyancy due to the high temperature of producer gas in the process of entrainment of air. Section 6.9 summarises the chapter and presents conclusions.

6.2   DOWNDRAFT BIOMASS GASIFIERS

Downdraft gasifiers are those in which the producer gas flows downwards along with the solids, and exits from the bottom end of the gasification reactor. Air for the gasification is introduced in the mid-height region of the gasifier. Open top gasification systems such as those designed by the Indian Institute of Science [13] permit 50–70% of the gasification air to enter from the top of the reactor, and the remaining enters from the *tuyeres*. There is also a third type called the stratified gasifier, where the air supply is only from the top, and no *tuyeres* are present [14]. Although this design was proposed, researched and analysed by the research group of Thomas Reed in the Solar Energy Research Institute (presently known as the National Renewable Energy Laboratory), no commercial versions of this type of gasifier were built till date. Mukunda [15] presents a detailed comparison of the different configurations and argues in favour of the open top reburn design. Figure 6.1 presents the schematic diagram showing the components of a downdraft biomass gasifier.

In the region of the *tuyeres*, since oxidation reactions happen predominantly, the reactor temperatures here are the maximum. Heat from this region is transferred to

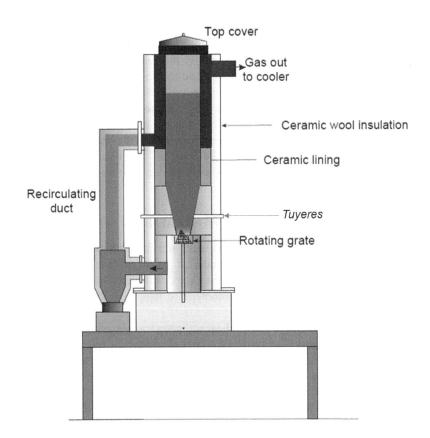

FIGURE 6.1   Schematic sketch of a typical downdraft biomass gasifier [13].

the biomass bed above the *tuyeres*, causing the biomass to devolatilise into char and volatiles, which in turn flow downwards through the oxidation region. Thus the bed below the *tuyeres* is primarily char, and the products of combustion (mainly CO_2 and H_2O) flow over this region and undergo reduction reactions to CO and H_2. Details of these phenomena are discussed in Section 6.3.

Need for a throat: Literature suggests that the oxidation region (*tuyere* region) should be at temperatures higher than 900–1000° C so that the volatile matter that passes through this region is completely cracked, and the exiting gas has no condensable matter, thus making the gas of a high quality, suitable for applications such as internal combustion engines and gas turbines [1], [2]. This requires that the cross-sectional area of the oxidation region is small, so that the oxidiser supply creates a high temperature region which occupies this entire cross-section. In order to facilitate this, there is a need for a converging or throat section. However, since the residence time of the reacting gases and solids needs to be adequate in the regions above and below the oxidation zone, there is a need for divergence of the reactor below the throat, and long reactor tube above the *tuyeres* [16], [15].

Design parameters of downdraft gasifiers: This earlier discussion indicates that the gasifier requires a long fuel bed above the *tuyeres*, and a convergent-divergent region below the *tuyeres*. The description of the factors affecting the required dimensions of these different functional regions of a gasifier is available in the literature, and is presented and discussed in Section 6.4. Three parameters are commonly used in the literature to quantify the gasification rates in a reactor vis-à-vis its size: The hearth load (B_h), defined as the rate of producer gas flow rate in Nm^3/h per cm^2 of the throat area of cross-section, is one such parameter. A variant of this is the Superficial Velocity (S_v), defined as the producer gas flow rate in Nm^3/s per m^2 of the throat area, thus making its units as (m/s) [17], [13]. A third parameter, defined by Kaupp and Goss [18] is the specific gasification rate, as the mass flow rate of biomass in kg/h per m^2 of the throat area. Most of the literature on gasifier design use one or more of these parameters in the design process.

6.3   PHYSICAL AND CHEMICAL PHENOMENA IN GASIFICATION

The foregoing discussion also refers to physical and chemical phenomena—viz., drying, devolatilisation, oxidation and reduction. In this section, these phenomena are discussed, with a view to identifying the desirable conditions the reactor needs to provide for each of these phenomena to happen in the most effective fashion.

Drying is predominantly a combined heat and mass transfer phenomenon, where heat transfer from the gases surrounding a biomass particle results in the rise in temperature of the particle and the moisture contained therein. The concentration of water vapour and the temperature and pressure of air surrounding the particle together determine the equilibrium moisture content of the particle. The difference between the actual moisture content of the particle and the equilibrium moisture content provides the driving force for the mass transfer phenomenon from the particle to the surroundings, resulting in drying.

Reed and Jantzen [17] suggest that the central part of the reactor cross-section receives heat by radiation from the oxidation region near the *tuyeres*, but the part near the walls in the fuel bed are relatively colder, resulting in the moisture and volatiles condensing on the walls and draining downwards. They recommend that there should be traps for capturing the condensate and draining it outside the reactor in order to prevent the condensate from flowing down to the reaction regions, cooling or quenching the hearth. They have also recommended that the walls be heated on the outside by the exiting producer gas to prevent/minimise the condensation. ABETS [13] also recommends a jacket of producer gas to surround the pyrolysis and drying regions in the fuel bed above the *tuyeres* in order to increase the hearth temperature and to prevent condensation of moisture and volatiles on the inner surface of the reactor walls.

Devolatilisation or pyrolysis occurs when biomass is heated to temperatures in excess of $200°$ C. The cellulose, hemicellulose and lignin contained in the biomass undergo thermal decomposition at elevated temperatures, releasing gases and volatiles, leaving behind char. Hemicellulose decomposes between $220–315°$ C, while cellulose pyrolyses between $315–400°$ C, and lignin breaks up over the range of $160–900°$ C. Heating rates in gasifiers are typically in the order of tens of degrees Celsius per minute [15]. Final products of devolatilisation depend on the temperature and heating rate, and are known to contain over 200 different compounds, some of them condensable and some of them non-condensable. The major constituents of non-condensables include CO, CO_2, H_2, H_2O, CH_4. Devolatilisation leaves behind a solid residue of char, which is known to be predominantly carbon, but also contains some hydrogen and oxygen in its constituents.

Oxidation reactions are the fastest and the most exothermic, and are localised to a small region downstream of the *tuyeres*, within which all oxygen available in the air is consumed. In this region, gaseous combustibles from devolatilisation—viz., H_2, CO, CH_4 and tar—are oxidised, besides surface oxidation of char. It has been observed that char oxidation is diffusion-limited, and results in shrinkage of char particles owing to the reaction being limited to diffusion of oxygen to the surface of char. On the other hand, reactions of H_2O and CO_2 with char are also kinetically limited owing to the reactions taking place at the solid surfaces inside the pores of char particles, due to diffusion of these gaseous species into the pores of char as well [15], [19]. Reduction reactions are heterogeneous reactions, occurring between gaseous and solid reactants at the interface, and are endothermic. Hence, they are also substantially slower than the oxidation reactions, and would need adequate residence of the reactants in the regions at high enough temperatures for the reactions to complete themselves.

Since in a downdraft gasifier, oxygen is supplied and consumed in a very small region, extent of oxygen consumption cannot serve as a measure of completeness of gasification: In fact, it is the reduction reactions that make combustible products of gasification. CO_2 and H_2O, on the other hand, are produced in devolatilisation as well as oxidation, and are consumed in the reduction region, and thus cannot be indicators of completeness of the reaction. Similarly CO and H_2 are also produced in the devolatilisation process and consumed during oxidation, and subsequently produced again in the reduction reactions. Thus, it is the extent of carbon conversion that can reliably be used as a measure of the extent of completion of gasification reaction [20].

6.4 REQUIREMENTS FROM A GASIFICATION REACTOR

6.4.1 DRYING REGION

The earlier discussion lists a few specific requirements of the various physical and chemical processes that the design of gasification reactor needs to take into account. Drying requires that the heat transfer from the hearth to the fresh feedstock by radiation through the bed, supported by conduction from a hot jacket around the reactor. In some designs, this is provided for by an annular passage for exiting hot producer gas around the upper part of the reactor above the *tuyeres*. It could also be achieved by insulating the reactor well. In combination, these two mechanisms of heat transfer should be able to remove as much moisture as possible before the feedstock enters the reaction zone. In case the feedstock contains more than 25% moisture, there would be a tendency for the moisture to find colder surfaces to condense and trickle down the bed into the hearth region. The condensate traps recommended by Reed and Jantzen [17] could prevent the condensate from affecting the reactions in the hearth region. Up to this limit of moisture content, it has been observed in all literature of downdraft biomass gasifiers that the feedstock that flows into the reaction region is adequately dried.

6.4.2 PYROLYSIS REGION

Devolatilisation is a phenomenon physically similar to drying, but the condensate is sticky and can cause bridging—a phenomenon where the particles of feedstock stick together owing to condensing volatiles, resulting in a solid barrier which stops further flow of the feedstock to the reaction zone, thus impairing the smoothness and continuity of gasification. A jacket of hot producer gas around the reactor or thick insulation helps prevent this occurrence as well. Since the volatile products of devolatilisation are corrosive, the reactor vessel in the pyrolysis region needs to be made out of corrosion-resistant stainless steel or ceramic [13]. Groenveld [16] presents heat transfer analysis to the feedstock particles above the *tuyeres* to compute the minimum residence time of particles in this region so as to enable them to be devolatilised completely before entering the hearth. Using this analysis, he tries to find justifications for some of the empirical recommendations on hearth design.

6.4.3 OXIDATION REGION

Oxidation region is usually small, and is confined to the region around the air inlet from the *tuyeres*. Reed [21] estimates the size of the oxidation zone in various gasifiers to be typically 2.66 times the particle diameter. Kaupp and Goss [18] and Reed and Das [14] explain the expected structure of the hot region and argue that unless the hot zones of all *tuyeres* merge to ensure no cold zone between them, there is likelihood of tar and unreacted larger particles passing through the cold zones escaping tar cracking, resulting in uneven gasification, increased tar in the gas and poor gas quality. ABETS [13] also emphasises that gasification itself must be tar-free, and attempts at recirculating the gas to crack tar can be complex, expensive but ineffective. For this reason, practically all available literature recommends a decrease in reactor cross-sectional area in the oxidation region just below the *tuyeres*.

Temperatures in this region quoted in various sources vary, but all agree that temperatures in excess of 900° C are essential. This necessitates that the material of construction of this zone should be high-temperature-resistant: ceramics or specialised furnace hearth construction material. ABETS [13] describes the various versions tried by the IISc research group, to ensure functionality as well as longevity of the system. The final version recommended by this reference is an outer mild steel shell, with one layer of cold-face insulation bricks and another of hot-face brick lining, and an innermost layer of ceramic tiles with 90% or higher alumina content for the reactor walls, and a special ceramic tube for *tuyeres*. Reed and Jantzen [17] and Reed and Das [14] have talked about replaceable components such as the hearth mantle and constriction ring made of cast iron. In case of damage to some parts that face the highest temperature and erosion, the entire hearth need not be re-fabricated, if these parts can be made separate and easily replaceable. Likewise, Kaupp and Goss [18] recommend a choke-plate, which has the shape of a plate with a central hole, made of ceramic: Its shape ensures minimal thermal stresses and hence a lower tendency for the ceramic plate to crack; secondly, it also provides for a more easily replaceable spare in the hottest part of the gasifier.

6.4.4 REDUCTION REGION

Length of the reduction region required for good carbon conversion has been estimated by various researchers. Mukunda [15] presents the relationship between particle diameter and time of residence required for the three main heterogeneous reactions: char-O_2, char-H_2O and char-CO_2. While the time for consumption of the particle shows dependence on the square of the particle size for oxygen, the powers are lower for water vapour and carbon dioxide. Since the d^2-dependence is characteristic of diffusion combustion of a particle, it can be expected that concentration of oxygen reaches zero at the surface of the solid particle. On the other hand, particle diameter powers lower than 2 indicates that the gaseous species has also diffused into the particle, and causes a kinetic dependence as well. Also, the time scales of consumption of char by H_2O at 1000° C are nearly same as that for char oxidation at the same temperature, while the time scale of reaction of char with CO_2 is an order of magnitude higher. This explains why the concentration of H_2O at gasifier exit can be significantly reduced by making the reduction zone longer, but the same is not the case with CO_2. In fact, char-CO_2 reaction rates increase at higher temperatures, which are unlikely in the reduction zone of a downdraft biomass gasifier. Littlewood [20] shows that at distances beyond 10–12 diameters of the char particle, the carbon consumption nearly saturates, and further increase in the length of reduction zone is likely to be less effective in improving gasification.

Need for long enough residence time in the reduction zone before the tempera-tures drop below about 700° C has been emphasised in most literature on downdraft gasifier systems. Just as there is need for convergence of the reactor for oxidation, in order to increase residence time in reduction zone, often, a divergent section is recommended, thus requiring a throat of minimum cross-section below the *tuyeres* and above the grate. Also, since char particles get consumed and become smaller, if the local gas velocities are too high, this results in dusting of the fine particles of char and ash into the gas stream, making carbon conversion incomplete and reducing efficiency of gasification. This also demands a diffuser region of increasing cross-sectional area. Another factor emphasised in the literature is the need for uniformity of temperature in the reduction region, since this region is the most critical in gasifi-cation. This is ensured by deceleration on one hand, and recirculations on the other, and a divergent section, depending on the angle of divergence, can result in both of these. Mukunda [15] suggests the use of open top reburn downdraft design in order to keep the superficial velocities lower even at large gasification rates, thus ensuring a good quality of gas over a larger range of gasification rates.

6.4.5   GRATE AND ASH REMOVAL

Another important factor to be taken into account in designing a gasification reac-tor is the removal of ash. In continuous operation, as the char is consumed in the reduction region, ash that is left behind accumulates. On one hand, this occupies the volume that would otherwise have been filled with char flowing down from above, and thus, reduces the effectiveness of gasification reactions. On the other, as the ash fills the high temperature regions and stays there for long, the tendency for ash fusion and clinker formation increases. If the ash has more potassium content, the ash fusion temperatures could be lower, and the tendency for ash fusion and clinker formation starts even at normal hearth temperatures [13]. Wood from most species has a small ash content, usually about 1% by dry mass. This means in wood gasifi-ers of capacity up to a few hundred kW, ash production is not in large quantities, and thus, periodic removal of ash would normally suffice, and manual shaking of the grate at regular intervals would be sufficient. However, in larger gasifiers, one might need to automate the process of ash removal, and it also needs to be continu-ous. Feedstock such as rice husk and paddy straw, on the other hand, have a much higher ash content: typically 20–30% by dry mass. Also, when agro-waste or pellets thereof are used as gasifier fuel, there is a tendency of these fuels to have a lot of soil/mud sticking to it during collection and transportation, and this results in increased ash content [13]. In such cases, the requirement for ash removal is more important. Thus, designing the grate which supports the fuel bed in downdraft gasifiers is of critical importance to the continuous, smooth functioning of the gasifier at desirable efficiency. Different designs of grates are available in the literature [17], [18].

6.5   REQUIREMENTS OF FUEL CHARACTERISTICS

Particle size: Drying and devolatilisation depend on the heat conduction through the particle and mass diffusion from the particle, both of which depend on the particle size, or in turn, the particle surface area per unit bed volume [13]. Too large a particle size slows down these phenomena, and can result in incompletely dried and/or devolatilised particle entering the hearth, and can result in poor gas quality and incomplete carbon conversion. Too small a particle size, on the other hand, can produce an excess of volatiles, which may not be completely cracked while passing through the hot reaction zone. Small fuel particle sizes also result in small char particles, which could get consumed faster and hence reduce the effective gasification in the reduction zone. Literature reports higher calorific value of producer gas as particle size is reduced, but the tar content in the gas increases. ABETS [13] and Mukunda [15] advocate the use of a mixed particle size in order to ensure a good rate of devolatilisation as well as adequate availability of char particles in the reduction zone to ensure good gasification and tar cracking. Recommended particle size range in the literature is between one-sixth to one-twelfth of the reactor diameter [13]. Kaupp and Goss [18] mention that the particles must be at least 6.8 times smaller than the reactor diameter to avoid blockade due to bridging and clogging. They also discuss another important issue that depends on the particle size: the pressure drop through the bed, and hence the power of the device required to push or suck air through the gasification reactor; both these quantities increase with decrease in particle size. This becomes an issue when the gasifier is connected to an engine that draws the air through the gasifier, since it directly affects the volumetric efficiency of the engine.

As the gasifier capacity decreases, the required throat sizes become smaller. For instance, the gasifiers designed by Sutar [10] had throat diameters of 20 mm and 30 mm. If the particle sizes are comparable to this are larger, there is a risk of blockade at the throat owing to particles getting stuck there, and this can affect smooth and continuous operation of the gasifier. The largest fuel particle size that could be used on these gasifiers were 15 mm and 18 mm respectively, and the best gasification efficiency was observed with particle sizes in the range of 10–12 mm.

Moisture content: Moisture content of the fuel also has interesting effects on the gasifier performance. When the fuel is oven-dried to less than 5% moisture, gas composition shows less hydrogen and hence a lower calorific value. Fuel consumption rate is also higher, leading to a lower efficiency of gasification. As the moisture content is increased to around 12%, the hydrogen content of the producer gas improves to give the maximum calorific value. When moisture content is increased further, the water vapour content in the producer gas increases, resulting in poorer calorific value. Also, more wet wood is consumed, resulting in poorer gasification efficiency [10]. ABETS [13] also terms this as a "double effect" of moisture content on gasification efficiency: (i) due to reduced biomass calorific value and (ii) due to changes in oxidation-reduction reactions in the presence of excess moisture. Sutar [10] reports an optimum moisture content in the range of 10–12% for maximum gasification efficiency.

Ash content and bulk density: ABETS [13] discusses the use of rice husk as received for gasification: three issues with this feedstock are low bulk density, high ash content and small particle size. While the low density fuel could be made to flow smoothly using a rotating grate, owing to the high ash content, the char does not continue to burn, but gets quenched immediately after the oxidation zone, and thus char conversion is poor: the ash contains a lot of unconverted char. Ash content in fuel was discussed in Section 6.4.5 with reference to grate design. Wood and Branch [22] mention that with ash content less than 6% by dry mass of feedstock, no slagging (ash melting) is encountered, but with 6–12% ash, severe slagging is observed. ABETS [13], however, says that it is possible to maintain reactor temperatures such that slagging of high-ash content rice husk can be completely prevented.

Bulk density (mass of fuel per unit volume of the fuel bed) is another important parameter. Most gasifiers designed for woody biomass rely on the flow of fuel by gravity, and this happens well when the bulk density of the fuel is higher than about 250–300 kg/m^3. When the density is lower than this range, special feeding arrangements may be necessary, and a general purpose woody biomass gasifier may not be able to handle such feedstock directly. There are several gasifier designs specifically for utilising rice husk [22] and municipal solid waste [23]. Wood and Branch [22] list desirable ranges of fuel properties for gasification feedstock. Besides the aforementioned parameters, this report also suggests a volatile content of at least 10%.

6.6   DESIGN OF DOWNDRAFT BIOMASS GASIFIERS

As is the case with most technologies, the biomass gasifier technology far predated the understanding of the science underlying its operation. The technology was in widespread use since 1850 in UK and 1920 in USA, when coal was gasified and piped to residences as "town gas" for lighting, heating and cooking purposes. During the Second World War (1939–45), in Europe, more than a million gasifiers were manufactured and used for fuelling automobiles, owing to the oil blockade during the war. In Sweden alone, more than 0.7 million gasifier-powered vehicles were in operation during this period [14]. A lot of literature is available from the pre-1950 period, documenting the technology as well as the difficulties of generating and using producer gas in automobile engines [17], [18]. But the first set of research investigations on the science of gasification started appearing in the 1970s, especially after the 1973 OPEC oil embargo, which caused a renewed interest in alternatives to oil for energy and transportation. Even in this, literature on coal gasification is more abundant than that on biomass gasification. Reed and Das [14] mention a bibliography of over 10000 publications that include all the literature mentioned previously, prior to 1988.

Literature available on the empirical design of downdraft biomass gasifiers, mostly of the closed-top Imbert type gasifier, have been summarised by Kaupp and Goss [18] and Reed and Das [14]. The unpublished work of Venselaar [24] greatly supplements the details of the design process and its link to the theoretical understanding. ABETS [13] and Mukunda [15] have worked further on these design approaches and contributed further to it, although they do not explicitly document a design procedure for the IISc design of gasifiers, which combine the advantages of the stratified downdraft gasifier proposed by Reed et al. [25] and the closed top Imbert gasifier described in the Swedish literature translated by Reed and Jantzen [17]. In this section, the literature available on the design of the various parts of the gasifier are presented in such a way as to help a designer use them for calculation of the dimensions of the various components, with particular reference to an Imbert

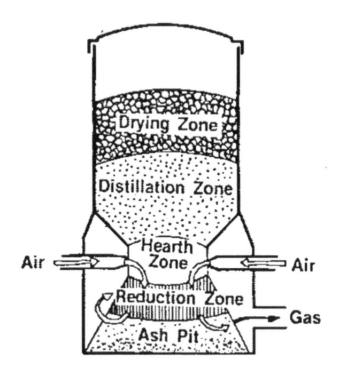

FIGURE 6.2   Imbert type downdraft biomass gasifier [14].

type gasifier shown in Figure 6.2. Experience of the author's group in extending this design process beyond the range of the data used in the design curves will also be discussed, along with literature where such attempts have been documented. Deviations from this design reported in the literature are also documented.

6.6.1   Fuel Container

The largest mass of the gasifier in the designs for transport application was the fuel drum, made of metal. Reed and Jantzen [17] recommended a volume of 100–200 litres for private cars, 200–300 litres for trucks and 300–600 litres for buses, for the fuel container. Since in the present day gasifiers are not used for transportation, there is flexibility about this volume, depending on the nature of operation and fuel consumption rates of the gasifier. In order to prevent/minimise condensation of moisture and volatiles on the colder surfaces of the fuel container, it is recommended [13] to use a double-walled structure, and allow the exiting high-temperature gases from the hearth to flow through the annulus. This serves two advantages: recovery of waste heat from the high-temperature gases to preheat the fuel feedstock on one hand, and prevention of condensation and avoiding the need for condensate collection traps. It is also suggested that the containers are concentric cylinders for ease of fabrication/erection. Material recommended for the inner wall of the container is stainless steel, since the volatile matter contains corrosive constituents such as acetic acid which can damage other metallic materials usually used for such fabrications. Reed and Jantzen [17] document that the use of aluminum coating was also not suitable owing to ammonium hydroxide content in the volatiles. The outer wall must be well-insulated from the exterior to prevent heat losses and accidental contact with hot surface by humans or other living beings. Recommended inner diameters of the container are discussed in Section 6.6.4.

6.6.2 THE HEARTH

This is the region where the oxidation and reduction reactions occur, and hence has the highest temperatures of the gasifier. Since the design requirements also demand a varying cross-section here—a convergent region for oxidation, where *tuyeres* are also present, a throat of minimum cross-sectional area and a divergent region at the bottom—this region is also expected to have the highest thermal stresses. Rapidly varying temperatures in the reaction zone from the inlet to the exit section of the hearth only adds to the complexity of this problem. This, along with possible non-axisymmetric temperature profiles expected in the lower parts of the hearth owing to change in flow direction of the gases at the exit, results in short lifespans of the components of the hearth [17]. The Swedish designers overcame this problem by introducing low-cost, replaceable components to constitute the V-hearth as shown in Figure 6.3: the mantle and the constriction ring. While the portions of these components which were not in the region of active gas flow got covered by a layer of ash that softens and sticks to the component, protecting the metal from overheating, those in the region of active flow got eroded, needing replacement after predefined life periods.

ABETS [13] re-designed the hearth as a continuously converging section, without a throat (see Figure 6.1). The material selection was also made so that the part of the hearth in direct contact with the highest temperatures—viz., the inner surface of the hearth and the *tuyere* tubes—could have a long enough life to justify major maintenance, and could be easily repaired or replaced. They chose ceramic tiles with more than 90% alumina content for the walls, and a special ceramic material for nozzle tubes. Tiles or tubes that look damaged during major maintenance can be replaced easily. ABETS [13] also recommends that the lower part consisting of the hearth of the gasifier be constructed using firebricks and ceramic tiles inside, and housed in a metallic outer container. The upper part comprising the fuel container could be separately fabricated and bolted with sealing to the lower part during assembly.

6.6.3 GASIFICATION AIR INLET

Gasification air can be taken into the hearth either in suction mode or in pressurised mode. Gasifiers connected to an engine system use the suction created by the engine to draw air into the reactor: the outlet of the gasifier is connected through a gas cleaning system to the engine, which creates a low pressure in the entire gasifier system, resulting in air being drawn through the *tuyeres* into the hearth. On the other

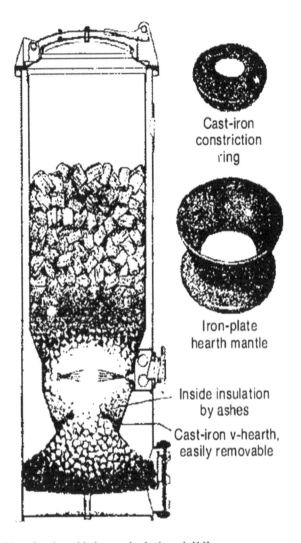

Cast-iron
constriction
ring

Iron-plate
hearth mantle

Inside insulation
by ashes

Cast-iron v-hearth,
easily removable

FIGURE 6.3   Use of replaceable inserts in the hearth [14].

hand, connecting a blower upstream of the *tuyeres* can push the air supply required by the gasifier through the *tuyeres*.

The documentation of Swedish experience [17], [18] presents multiple alternatives for air supply to the hearth: (i) single central pipe from above, opening into the hearth; (ii) single central pipe from below, opening into the hearth through radial holes, and (iii) multiple *tuyeres* opening through the walls of the reactor into the hearth. Out of these, the last one has been recommended as the most suitable for Imbert gasifiers, and has been used extensively. Several variants of the air supply to *tuyeres* have been discussed in these documents. Out of these, one with a single air inlet into a volume around the hearth, with air entering the hearth through the holes provided in a cast iron mantle, has been recommended to make the airflow through all *tuyeres* nearly equal. ABETS [13] on the other hand has recommended ceramic tubes embedded into the insulation brick walls appropriately to form the *tuyeres*. The size and number of *tuyeres* required are discussed in Section 6.6.4.

6.6.4 SIZING A GASIFIER

In this section, a procedure for sizing Imbert type downdraft biomass gasifiers using air as oxidising medium is discussed. It is known that the lower heating value of producer gas from such a gasifier is in the range of 3–5 MJ/Nm³. The steps in sizing of a gasifier are as follows:

1. The maximum thermal power output expected from a gasifier for its intended application is the starting point of sizing of a gasifier. In case the gasifier is intended to be used to run an IC engine or a gas turbine, the typical efficiency of such a prime mover at maximum power is used to compute the thermal power input needed for the prime mover. Let the intended output maximum thermal power of the gasifier be denoted as $P_{max,th}$. Since the typical heating value of producer gas is known, and denoted as LHV, this can be converted into the required maximum generation rate of dry producer gas $\dot{m}_{pg,max}$, using the expression

$$\dot{m}_{pg,max} = \frac{P_{max,th}}{LHV}.\qquad\qquad 6.1$$

Once the maximum required producer gas flow rate is known, the consumption rates of biomass and air can be estimated using the typical range of values of air-biomass ratio. Typically, biomass has a stoichiometric air-biomass ratio of around 6, and downdraft gasifiers are known to perform best in the range of 0.25–0.4 times the stoichiometric air-biomass ratio (i.e., the typical air-biomass ratio would be in the range of 1.5 to 2.4). This would help in calculating the maximum consumption rate of dry biomass $\dot{m}_{db,max}$ and maximum dry airflow rate $\dot{m}_{da,max}$ for which the system is to be designed. If we denote the range of air-biomass ratio on dry basis as $\left[ABR_{min}, ABR_{max}\right]$ and ash mass fraction in biomass as $X_{ash,d}$ on dry basis, then we get

$$\dot{m}_{db,max} = \frac{\dot{m}_{pg,max}}{ABR_{min} + 1 - X_{ash,d}}\qquad\qquad 6.2$$

$$\dot{m}_{da,max} = \frac{\dot{m}_{pg,max}}{1 + \dfrac{1 - X_{ash,d}}{ABR_{max}}}.\qquad\qquad 6.3$$

It should be observed here that the range 0.25–0.4 mentioned previously is the actual stoichiometric air-biomass ratio, and is customarily referred to as *equivalence ratio* in gasification literature. Strictly speaking, in combustion literature, the term equivalence ratio refers to the reciprocal of this quantity. One needs to pay attention to this discrepancy of terminology. In the present chapter, the use of the term *equivalence ratio* is avoided, and instead, explicit definition of the quantity is used to avoid ambiguity.

If the moisture fraction in biomass by mass on as-received basis is X_m and the specific humidity of the ambient air supplied to the gasifier is ω_a, the maximum consumption of wet biomass (as received) and moist air (atmospheric air) can be calculated using

$$\dot{m}_{b,max} = \frac{\dot{m}_{db,max}}{1 - X_m} \qquad 6.4$$

$$\dot{m}_{a,max} = \dot{m}_{da,max}\left(1 - \omega_a\right). \qquad 6.5$$

2. Once the maximum mass flow rates of producer gas, air and biomass are calculated, the recommended range of parameters available in the literature can be used to compute the throat diameter of the gasifier. Most other dimensions of the gasifier are related to the throat diameter, and can be computed once the throat diameter is computed.

Throat diameter D_{th} (in meters) is computed using one of the following parameters recommended in the literature: hearth load B_h, Superficial Velocity S_v or Specific Gasification Rate (*SGR*) defined in Section 6.2:

$$D_{th} = 0.677\sqrt{\frac{\dot{m}_{pg,max}}{\rho_{pg}^o B_h}} = 1.1284\sqrt{\frac{\dot{m}_{pg,max}}{\rho_{pg}^o S_v}} = 67.7\sqrt{\frac{\dot{m}_{b,max}}{SGR}}. \qquad 6.6$$

In this equation, ρ_{pg}^o stands for the density of producer gas at 1 bar, 0° C, the conditions at which Nm^3 is defined. In order to determine this, a typical producer gas composition needs to be assumed. Since typical density of producer gas is reasonably close to that of atmospheric air owing to close proximity of their molecular weights, one may not be in major error if the density of air at 1 bar, 0° C (i.e., 1.2763 kg/m^3), is used for ρ_{pg}^o. This would reduce equation (6.6) to

$$D_{th} = 0.6\sqrt{\frac{\dot{m}_{pg,max}}{B_h}} = \sqrt{\frac{\dot{m}_{pg,max}}{S_v}} = 67.7\sqrt{\frac{\dot{m}_{b,max}}{SGR}}. \qquad 6.7$$

Recommended ranges of values are available in the literature for B_h, S_v and *SGR* in the aforementioned equations. Reed and Jantzen [17] and Kaupp and Goss [18] specify the range of values recommended for Hearth Load B_h as 0.3–1.0 Nm^3/h/cm^2, and that for Specific Gasification Rate *SGR* as 600–3900 kg/m^2-h, although they cite a few exceptions of gasifiers operating outside this range.

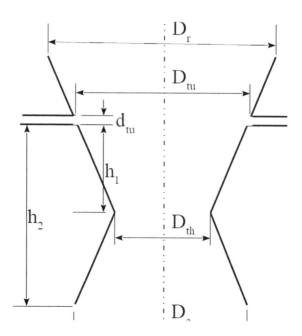

FIGURE 6.4 Schematic showing the characteristic dimensions of the gasifier hearth.

3. The other dimensions of the hearth and gasifier are recommended by Kaupp and Goss [18] based on the sketch of a gasifier hearth given in Figure 6.4. The ratios of the various dimensions—viz., the reactor diameter D_r, diameter of the circle marking the tip of the *tuyeres* D_{tu}, the elevation of the *tuyeres* above the throat h_1 and the elevation of the *tuyeres* above the exit section of the gasifier h_2 with the throat diameter D_{th}—are recommended for Imbert gasifiers based on the Swedish experience by Reed and Jantzen [17] and summarised by Kaupp and Goss [18]. Sutar [10] plotted these parameters and extrapolated them non-linearly to obtain dimensions of gasifier hearths smaller than those available in the earlier references. These plots are reproduced in Figure 6.5, Figure 6.6 and Figure 6.7.

Corresponding to the diameter of the throat of the hearth D_{th} calculated as described by equations (6.6) or (6.7), the diameters of the reactor (D_r) and *tuyeres* circle (D_{tu}) can be obtained from Figure 6.5. The exit diameter of the divergent region D_e is recommended to be $2.5D_{th}$ by Reed and Jantzen [17]. Likewise, the length of the divergent region, denoted by $\left(h_2 - h_1\right)$ as per the dimensions depicted in Figure 6.4, is recommended to be equal to the throat diameter D_{th}. The maximum distance of the throat below the *tuyeres* is given by Figure 6.6, while the diameter of the *tuyeres* is as per Figure 6.7. In this figure, the number of *tuyeres* N_{tu} is chosen such that the velocity of air through the *tuyeres* typically falls in the range of 25–35 m/s [17].

The length of the reactor above the converging section of the hearth can be chosen depending on how much fuel one would like to hold in the gasifier in one feed,

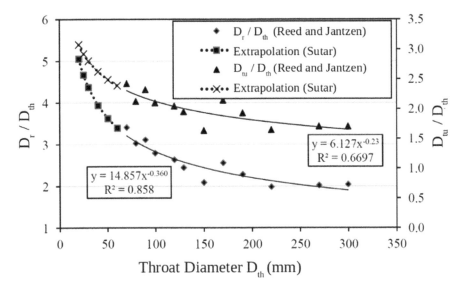

FIGURE 6.5   Recommended diameters of *tuyeres* circle and reactor [17], [10].

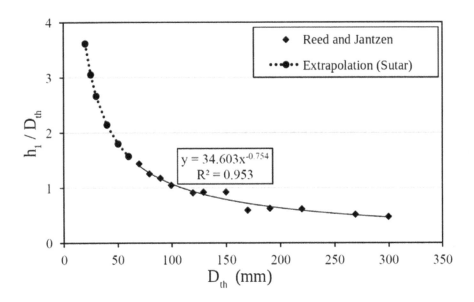

FIGURE 6.6   Recommended height of *tuyeres* above the throat [17], [10].

so that the gasifier can operate uninterrupted for a reasonable length of time between feeds of fuel. This is more critical for the Imbert type gasifiers which have a closed airtight top lid, which should not be opened during the functioning of a gasifier. In the open top reburn designs, the top of the gasifier is open by design, and fuel feeding during operation of a gasifier is not a problem with such designs. Considerations

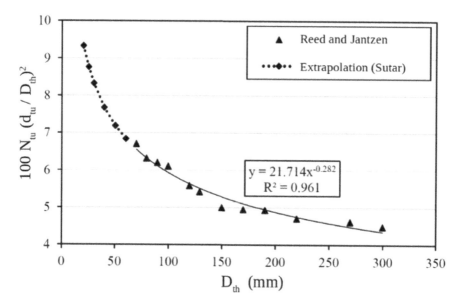

FIGURE 6.7   Recommended diameter of *tuyeres* [17], [10].

on the volume of the fuel container have already been discussed in Section 6.6.1, which could be used in conjunction with the calculated value of reactor diameter D_r to arrive at a suitable value of the length of the fuel container.

6.7   DESIGN PROCEDURE FOR AERATED BURNERS FOR BOTTLED GASEOUS FUELS

Completely or partially aerated burners are commonly used for burning bottled gaseous fuels such as natural gas, LPG, etc. both in the industrial sector and in the domestic sector for heating applications. Jones [7] comprehensively documents design procedures for such burners. This section summarises the procedure outlined by Jones [7], which forms the basis for the design procedure outlined in the next section for hot producer gas burners. In naturally aerated burners, the momentum/ flow energy of the moving gas stream is commonly used to entrain atmospheric air. The burner design process ensures that appropriate proportion of air is entrained at all flow rates of the gas in the design range so that a combustible mixture is produced and a stable flame is anchored at the burner. The procedure outlined by Jones [7] deals with bottled gases, which emerge from a nozzle with a certain velocity. Entrainment of the surrounding air into this turbulent jet is accounted for using conservation equations of momentum, flow energy and mass to derive the expressions for the required areas/dimensions of different salient components of the burner.

Figure 6.8 shows a typical aerated burner whose design process is detailed by Jones [7]. A nozzle connected to the fuel source through a regulator ensures a constant flow rate of the combustible gas in the form of a jet. Through momentum

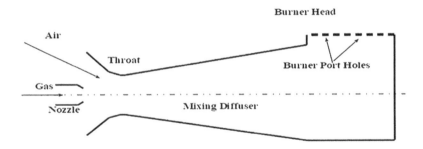

FIGURE 6.8 Schematic showing the characteristic dimensions of a naturally aerated burner.

transfer to the surrounding air and turbulent mixing, primary air is entrained into the stream, upstream of the throat. The fuel and air mix in the divergent (diffuser) part of the burner, and the mixture exits through the ports on the burner head. All these parts are clearly labelled in the figure.

6.7.1 DERIVING RELATIONS FOR AREA RATIOS

For partially aerated burners of hydrocarbon fuels, it is typical to use 60% of stoichiometric requirement of air as primary air, allowing the burner head to entrain the remaining air needed to complete the combustion. The diameter of the nozzle ensures a certain kinetic energy of the gas stream. Using the flow energy equation for frictionless flow and correcting it using discharge coefficient to account for flow losses in the nozzle, the volumetric flow rate of fuel gas $\dot{Q}_g$ can be written as

$$\dot{Q}_g = C_d A_j \sqrt{\frac{2\Delta p}{\rho_g}} \qquad\qquad 6.8$$

where the flow is assumed incompressible, C_d is the discharge coefficient of the nozzle, A_j is the area of the nozzle for gas flow, Δp is the pressure difference across the nozzle and ρ_g is the density of the gas. Jones [7] recommends the length-to-diameter ratio of the hole of the nozzle to be close to unity, and the angle of convergence of the nozzle upstream of this hole to be about 35° so as to keep the discharge coefficient C_d higher than 0.9 (to reduce losses at the nozzle). The Reynolds number of flow through the nozzle hole is defined as $Re_d = \rho_g V_{gd}/\mu_g$ where d is the diameter of the hole of the nozzle, V_g is the velocity of gas flow through it and $\propto_g$ is the coefficient of viscosity of the gas. It should be noted that $A_j = \pi d^2/4$ and $V_g = \dot{Q}_g / A_j$. It is recommended that the Reynolds number for the nozzle holes be maintained around 5000 or slightly higher for the entire range of flow rates of nozzle operation, so that the discharge coefficient remains reasonably constant over the operating range of the burner. Operation at higher Reynolds numbers could cause a whistling noise at the orifice, and at lower Reynolds numbers the discharge coefficient decreases rapidly, both of which should be avoided.

Entrainment of air into the jet is calculated using the momentum and flow energy equations for flow through the burner. The momentum equation between the inlet of the venturi (convergent part) of the burner, which is also the location of the nozzle, and the throat can be written as

$$\left(P_t - P_o\right)A_t = \rho_g V_g^2 A_j - \rho_m V_m^2 A_t = \frac{\rho_g \dot{Q}_g^2}{A_j} - \frac{\rho_m \dot{Q}_m^2}{A_t}$$
6.9

where subscripts t and o stand for throat and venturi inlet respectively, and subscripts g and m denote the fuel gas coming from the nozzle and gas-air mixture at the throat of the venturi respectively. Similarly, the flow energy equation between the throat and burner port inlet (just below the burner ports, inside the burner) can be written as

$$\left(P_p - P_t\right) = \frac{\rho_m V_m^2}{2}\left(1 - C_L\right) = \frac{\rho_m \dot{Q}_m^2}{2A_t^2}\left(1 - C_L\right).$$
6.10

Here C_L is the loss coefficient for the mixing diffuser region, defined as the ratio of pressure loss in this region to the mixture kinetic energy at the throat of the venturi. In both equations, $\dot{Q}_m$ is the flow rate of the mixture (m³/s) and V_m is its velocity (m/s) at the throat section. Eliminating p_t between equations (6.9) and (6.10), we get

$$\left(P_p - P_o\right) = \frac{\rho_g \dot{Q}_g^2}{A_j A_t} - \frac{\rho_m \dot{Q}_m^2}{2A_t^2}\left(1 + C_L\right).$$
6.11

Jones [7] defines a dimensionless pressure efficiency η as the ratio of $P_p - P_o$ to the kinetic energy of the jet leaving the injector as

$$\eta = \frac{P_p - P_o}{\frac{\rho_g V_g^2}{2}}.$$
6.12

Using equations (6.11) and (6.12), he expressed the pressure efficiency η in terms of the area ratio A_j / A_t, and maximises η by differentiating it with respect to this area ratio, to obtain the value of this ratio for minimal losses through the system. This gives rise to the expression

$$\frac{A_j}{A_t} = \frac{\rho_g \dot{Q}_g^2}{\rho_m \dot{Q}_m^2\left(1 + C_L\right)}.$$
6.13

Defining entrained air-to-gas volume ratio $R = \dot{Q}_a / \dot{Q}_g$, we can write

$$\dot{Q}_m = \dot{Q}_g\left(1 + R\right).$$
6.14

If we define the relative density of the gas with reference to air as $\sigma = \rho_g / \rho_a$, then using the mass balance for the mixing process $\rho_m \dot{Q}_m = \rho_g \dot{Q}_g + \rho_a \dot{Q}_a$, we can define the density ratio of mixture-to-gas as

$$\rho_m / \rho_g = (\sigma + R) / \sigma (1+R).$$ 6.15

Using these parameters, equation (6.13) can be rewritten as

$$\frac{A_j}{A_t} = \frac{\sigma}{(\sigma+R)(1+R)(1+C_L)}.$$ 6.16

The ratio of burner ports to throat areas is also obtained from these previous equations, by substituting for $\dot{Q}_m$ and ρ_m / ρ_g respectively from equations (6.14) and (6.15) in equation (6.11):

$$P_p - P_o = \frac{\rho_g \dot{Q}_g^2}{A_t^2} \left(\frac{A_t}{A_j} - \frac{(\sigma+R)(1+R)(1+C_L)}{2\sigma} \right).$$ 6.17

In equation (6.17), if we substitute for A_t / A_j from equation (6.16) for an optimised system, we get

$$P_p - P_o = \frac{\rho_g \dot{Q}_g^2}{A_t^2} \frac{(\sigma+R)(1+R)(1+C_L)}{2\sigma}.$$ 6.18

For flow of the gaseous mixture across the burner ports, we could write the flow energy equation, using a discharge coefficient C_{dp}. In writing this, we recognise that at the top end of the burner ports in Figure 6.8, the pressure is ambient (p_o) again:

$$P_p - P_o = \frac{\rho_m \dot{Q}_m^2}{2A_p^2 C_{dp}^2}$$ 6.19

where A_p is the total area of the burner ports. Here again, substituting for ρ_m / ρ_g and $\dot{Q}_m / \dot{Q}_g$ as already mentioned, we get

$$P_p - P_o = \frac{\rho_g \dot{Q}_g^2}{A_p^2} \frac{(\sigma+R)(1+R)}{2\sigma C_{dp}^2}.$$ 6.20

Equating the right hand sides of equations (6.18) and (6.20) we get

$$\frac{A_t}{A_p} = C_{dp} \sqrt{(1+C_L)}.$$ 6.21

It is interesting to see that this equation is independent of the gas or mixture flow rates or properties, except indirect dependence through the discharge coefficient C_{dp} and loss coefficient C_L. In the frictionless ideal flow case where $C_{dp} = 1$ and $C_L = 0$, we would get $A_p = A_t$. The ratio of injector area to ports area can be obtained by combining equations (6.16) and (6.21) as

$$\frac{A_j}{A_p} = \frac{\sigma C_{dp}}{(\sigma + R)(1 + R)\sqrt{(1 + C_L)}}.$$ 6.22

6.7.2 PROCEDURE OF DESIGN CALCULATIONS

Now that we have derived all required relations, we should now be in a position to compute the relevant dimensions of a burner required for a given application, using the following steps.

1. For a given application and fuel, the power requirement $\dot{P}$ decides the gas flow rate $\dot{Q}_g$:

$$\dot{P} = \eta_c \rho_g \dot{Q}_g LHV$$ 6.23

   where η_c is the combustion efficiency, defined as the fraction of energy contained in the fuel that is actually released by combustion. For most applications, this parameter could be assumed to be unity, since in most well-designed burners, hardly any fuel escapes unburned.

2. The pressure of gas upstream of the nozzle, as permitted by the regulator, decides the pressure difference across the nozzle $\mathscr{D}p$. If the geometry of the nozzle is known, its discharge coefficient C_d can be obtained from literature. Then, using equation (6.8), the value of nozzle area A_j can be calculated.

3. Once the fuel is known, its relative density A under ambient conditions is known. Typically, a primary aeration of 60% of stoichiometric air requirement is recommended for partially aerated burners, and 100% or more than stoichiometric requirement of combustion air for totally aerated burners. Using this, the air-to-gas ratio $R = \dot{Q}_a / \dot{Q}_g$ can be evaluated. Then using equation (6.16), the required throat area can be calculated. The diffuser loss coefficient C_L in this equation takes values between 0.25–0.35 for lengths of the convergent and divergent portions of the burner venturi equal to 2.5 and 10 times the throat diameter respectively, and a convergence angle of about 35–40° and a divergence angle of about 10–14° [7].

4. The ratio of throat to port areas is given by equation (6.21). For typical port geometries, the range of C_{dp} values are suggested to be 0.6–0.7 [7]. Using this in equation (6.21), the ratio of throat to port areas can be determined. Knowing the throat area from step 3, the total area of the ports can be calculated.

5. The diameter of each hole of the burner ports should be evaluated depending on the quench distance of the fuel. The diameter of each hole must be smaller than the quench distance, and the length of each hole should typically be at least twice the diameter of the hole, so that in case of flashback, the flame does not enter the burner through the hole but gets

quenched inside it. Jones (1989) states that the quenching diameter of the hole is related to the thermal diffusivity and burning velocity of the fuel as $d_o = k\alpha / S_u$ where α is the thermal diffusivity of the unburned mixture, S_u is the burning velocity of the mixture and k is a constant. According to Jones [7], the limiting port diameter for a flame of hydrogen is the smallest, at 0.8 mm, while that for methane is 3.5 mm, and that for most other hydrocarbon fuels lies in between. Using the area of one such port hole, and the total area A_p, the number of holes on the burner head can be determined. The cross check is that the velocity of the mixture leaving each port hole should be able to anchor a small conical flamelet at the port hole, whose semi-angle at the vertex, α_f, is related to the burning velocity and flow velocity as $S_u = V_m \sin\alpha_f$, and the angle α_f should be an acute angle in the range of 15–45°: this means that the mean mixture velocity V_m through each burner port hole is such that $V_m > S_u$, and the cone angle is in the previously mentioned range.

6.7.3   ILLUSTRATIVE EXAMPLE CALCULATION

In order to illustrate the procedure, sample calculations for a domestic LPG cookstove burner are shown here. The burner is designed for a power output of 1367 kcal/h (1.59 kW_{th}), or 125 g/h of LPG, typically 70% butane and 30% propane. Using this information, and taking the discharge coefficient of the nozzle C_d to be 0.85, we can calculate the nozzle area A_j from equation (6.8) as 0.361 mm² (diameter around 0.7 mm). In doing this, the density of the gas of the composition at 1 bar pressure and 30° C, evaluated from the ideal gas equation of state to be $\rho_g = 2.136$ kg/m^3, has been used.

Stoichiometric air requirement for the previously mentioned LPG can be evaluated using the following chemical equation for its complete combustion:

$$0.7C_4H_{10} + 0.3C_3H_8 + 6.05\left(O_2 + \frac{79}{21}N_2\right) \rightarrow 3.7CO_2 + 4.7H_2O + 6.05\left(\frac{79}{21}\right)N_2.$$

For this chemical equation for stoichiometric combustion of LPG, primary aeration of 60% can be evaluated to be 28.8 moles of air per mole of LPG. Using this, the value of $R = \dot{Q}_a / \dot{Q}_g$ comes out to be 17.28. The relative density of the LPG can be evaluated to be $\sigma = 1.862$ from the molecular masses of LPG (53.8) and air (28.9). Substituting these and using $C_L = 0.3$ in equation (6.16), the throat area can be calculated to be 88.21 mm² (throat diameter 10.6 mm).

Using $C_{dp} = 0.6$ in equation (6.21), the total port area can be computed as 128.94 mm². Using the recommended port hole diameter of 2.5 mm for LPG, the area of each hole of the burner port comes out to be 4.91 mm², making the number of such holes required on the burner port to 27. The results are summarised in Table 6.1, along with calculated results for a burner of the same power output (1.59 kW_{th}) that uses methane as fuel.

TABLE 6.1
Calculated Sample Results for a Domestic Burner with 1.59 kW$_{th}$ Output.

Fuel Gas	Mol. wt	σ	R	A_j mm² (d_j mm)	A_t mm² (d_t mm)	A_P mm²	d_h mm	A_{holes}
LPG	53.18	1.8616	17.28	0.361 (0.7)	88.21 (10.6)	128.94	2.5	27
Methane	16.0	0.5536	5.712	0.778 (1.0)	76.81 (9.9)	112.28	3.0	16

6.8 BURNER DESIGN FOR HOT PRODUCER GAS

When we set out to design a burner to burn producer gas emanating from a gasifier at high temperature, there are several factors that need to be taken into account:

1. The calorific value of producer gas per unit volume is substantially lower than that of typical hydrocarbon fuels. As compared to LPG, whose calorific value is between 90 and 95 MJ/Nm³, producer gas has a calorific value of around 4 to 5 MJ/Nm³. Since the temperature at which gas is available at burner inlet is close to 500 K in the case of downdraft gasifiers, the calorific value per m³ at this temperature is only about half this value. Thus, volumetric flow rate of producer gas required for the same output power is nearly 20–40 times that of LPG, depending on the temperature of producer gas.
2. The stoichiometric molar air-fuel ratio of LPG was seen in Section 6.7.3 to be 28.8, and thus the primary aeration needs to be around 17–18 times the volumetric flow rate of LPG. On the other hand, stoichiometric molar air-fuel ratio for producer gas is close to unity. Thus, the amount of primary air required to be entrained per unit volume of producer gas is close to 0.6–0.7 times the gas flow rate. However, since the gas flow rate of producer gas for a given power is 20–40 times LPG flow rate, the primary aeration required for a given power output is 12–28 times the volumetric flow rate of LPG, which is comparable to the primary aeration required for LPG or slightly more.
3. Since bottled gases are available at pressures substantially higher than atmospheric, pressure regulators are used to reduce this to, for example, 2 to 3 kPa gauge for supplying to domestic burners. Thus the ∅p in equation (6.8) is dictated by the regulator. In the case of producer gas, a blower is typically used to push the gas through the reactor or extract gas from it, and thus the pressure of gas at the burner is dictated by the blower. These pressures are usually much lower, in the range of 0.1–0.5 kPa gauge. Thus the gas velocities at the exit of the nozzle are substantially lower in the case of producer gas. This, combined with large volumetric flow rate requirements, results in A_j values substantially larger than those encountered in the case of bottled gaseous fuels.
4. While these factors could be taken care of by the calculation procedure outlined by Jones [7], the presence of buoyancy in the gas owing to its high

temperature is not accounted for by this procedure. In entraining primary air, in addition to the momentum of the gas flow, buoyancy also contributes substantially, and this needs to be taken into account for correctly designing burners for producer gas.

In this section, the procedure outlined by Sutar [10] and Sutar et al. [12] is presented and discussed. Reviewing the procedure outlined in Section 6.7.2, we find that it is only step 3 of the procedure that needs modification in order to account for buoyancy, since equation (6.16) needs to be replaced with a relation derived for buoyant fuel jets. The equation (6.21) applies equally well to burners for buoyant fuel jets as well, so long as the burner configurations are not very different. Also, the procedure for calculating the hole size on the burner port and the number of holes on the port outlined in step 5 remains unaltered.

A buoyant vertical fuel jet emanating from the nozzle of the burner has contributions to entrainment of surrounding air both from the velocity (inertia) of the jet as well as from the difference in density between the fuel jet and its surroundings. As we move along the jet axis, we find that owing to transfer of momentum to the stationary surrounding air, the flow decelerates, thereby decreasing the contribution of the flow velocity to further entrainment. Similarly, mixing of surrounding air with the hot fuel jet also lowers its temperature, thereby decreasing its potential for further entrainment. However, these two phenomena occur over different length scales [26]. Following Sutar [10] and Sutar et al. [12], locating the throat of the burner with reference to the nozzle could be most effective when the following two conditions are simultaneously ensured:

1. The strengths of the buoyancy and inertial components are similar or equal.
2. The diameter of the jet at the location of the throat is equal to or slightly smaller than the diameter of the throat.

The first of these two conditions is used to calculate the axial distance of the throat above the nozzle outlet, while the second is used to calculate the throat diameter in relation to the nozzle diameter.

Following Lee and Chu [26], in a turbulent buoyant jet, the initial part of the jet is usually momentum-dominated or inertia-dominated (see Figure 6.9). This is followed by a region where the two mechanisms have similar strengths, and further downstream, the inertia effects are weaker, and the mechanism of entrainment is buoyancy-dominated. Lee and Chu [26] define a momentum length scale l_s in terms of the nozzle-exit momentum flux $MO_o = \rho_g \dot{Q}_g V_g$ and buoyancy flux $F_o = (\rho_a - \rho_g)\dot{Q}_g g$ as

$$l_s = \frac{\left(\dfrac{MO_o}{\rho_g}\right)^{3/4}}{\left(\dfrac{F_o}{\rho_g}\right)^{1/2}}. \qquad\qquad 6.24$$

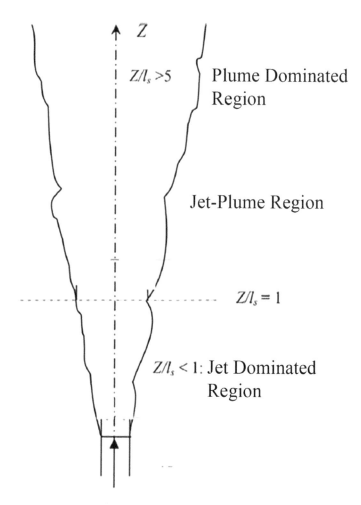

FIGURE 6.9 Regimes in a jet-plume [12], [26].

According to Lee and Chu [26], when $z < l_s$, the regime is inertia-dominated, and when $z > 5l_s$, the regime is buoyancy-dominated. Between these two limits, the two mechanisms of entrainment have similar strengths. Using the first condition stated earlier, if the location of the burner throat above the nozzle, z_t, is kept equal to l_s, the two mechanisms have similar strengths at the throat. Thus,

$$z_t = l_s = \frac{\left(\dfrac{MO_o}{\rho_g} \right)^{3/4}}{\left(\dfrac{F_o}{\rho_g} \right)^{1/2}}. \qquad\qquad 6.25$$

Again, following Lee and Chu [26], the plume width increases with z linearly, as $d(z) = d + 0.21z$, where d is the diameter of the fuel nozzle. The coordinate z is chosen such that $z = 0$ at the nozzle exit. $d(z)$ is the diameter of the jet at any axial distance z from the nozzle. Using this relation and the second condition stated earlier, the diameter of the throat can be written as

$$d_t = d + 0.21z_t. \qquad\qquad 6.26$$

The ratio of areas A_t / A_j is thus the ratio d_t^2 / d^2, and this can be used to replace equation (6.16), as:

$$\frac{A_t}{A_j} = \left(1 + 0.21\frac{z_t}{d}\right)^2. \qquad\qquad 6.27$$

Once this is done, the same procedure outlined in Section 6.7.2 can be used to design the burner for a buoyant fuel jet, with the exception that equation (6.27) is used in place of equation (6.16). Sutar [10] and Sutar et al. [12] have shown by modelling and measurements that a burner so designed also works with a primary aeration close to 60% of stoichiometric air, for hot producer gas emerging from a small downdraft gasifier at a temperature of around 500 K.

6.9   SUMMARY AND CONCLUSIONS

This chapter presented methods of designing downdraft biomass gasifiers and naturally aspirated burners for producer gas at high temperatures. The documented information of Swedish experience with designing and operating biomass gasifiers has been reviewed and combined with more recent experiences in presenting the design methodology for downdraft biomass gasifiers. The method of designing burners for high temperature producer gas has been derived based on the method available in the literature for bottled hydrocarbon gases, and has been derived by accounting for buoyancy as an additional mechanism of entraining combustion air. The experience of the author's research group in the areas has been integrated with the review of available information in the literature, in order to make a compilation that is directly useful to the researcher and practitioner.

REFERENCES

[1] K. Engvall, T. Liliedahl and E. Dahlquist, "Biomass and black liquor gasification," in *Technologies for converting biomass to useful energy*, CRC Press, 2013, pp. 175–216.
[2] P. Basu, *Biomass gasification and pyrolysis: Practical design and theory*, Academic Press, 2010, Burlington USA.
[3] C. Higman and M. van der Burgt, *Gasification*, Gulf Professional Publishing, 2003, Texas USA.
[4] S. Dasappa, U. Shrinivasa, B. N. Baliga and H. S. Mukunda, "Five-kilowatt wood gasifier technology: Evolution and field experience," *Sadhana*, vol. 14, no. 3, pp. 187–212, 1989.
[5] H. S. Mukunda, S. Dasappa, P. J. Paul, N. K. S. Rajan and U. Shrinivasa, "Gasifiers and combustors for biomass-technology and field studies," *Energy for Sustainable Development*, vol. 1, no. 3, pp. 27–38, 1994.

[6] H. S. Mukunda, S. Dasappa, P. J. Paul, N. K. S. Rajan, M. Yagnaraman, D. R. Kumar and M. Deogaonkar, "Gasifier stoves-science, technology and field outreach," *Current Science*, vol. 98, no. 5, p. 00113891, 2010.

[7] H. R. N. Jones, *The application of combustion principles to domestic gas burner design*, Taylor & Francis, 1989, Oxfordshire.

[8] P. Bhoi and S. Channiwala, "Optimization of producer gas fired premixed burner," *Renewable Energy*, vol. 33, no. 6, pp. 1209–1219, 2008.

[9] N. Panwar, B. Salvi and V. S. Reddy, "Performance evaluation of producer gas burner for industrial application," *Biomass and Bioenergy*, vol. 35, no. 3, pp. 1373–1377, 2011.

[10] K. B. Sutar, *Development of downdraft gasifier cookstoves for domestic application*, Indian Institute of Technology Delhi, New Delhi, 2015.

[11] K. B. Sutar, S. Kohli and M. R. Ravi, "Design, development and testing of small downdraft gasifiers for domestic cookstoves," *Energy*, vol. 124, pp. 447–460, 2017.

[12] K. B. Sutar, M. R. Ravi and S. Kohli, "Design of a partially aerated naturally aspirated burner for producer gas," *Energy*, vol. 116, pp. 773–785, 2016.

[13] ABETS, *Biomass to energy: The science and technology of the IISc bio-energy systems*, Indian Institute of Science, Bangalore, 2003.

[14] T. B. Reed and A. Das, *Handbook of biomass downdraft gasifier engine systems*, Biomass Energy Foundation, 1988, Colorado USA.

[15] H. S. Mukunda, *Understanding clean energy and fuels from biomass*, Wiley India, New Delhi, 2011.

[16] M. Groeneveld, *The co-current moving bed gasifier*, TU Delft, The Netherlands, 1980, Delft, Netherland.

[17] T. Reed and D. Jantzen, *Generator gas: The Swedish experience from 1939–1945 (a translation of the Swedish book, Gengas)*, Solar Energy Research Institute, Golden, CO, 1979, Delft, Netherland.

[18] A. Kaupp and J. R. Goss, *State of the art report for small scale (to 50 kw) gas producer-engine systems*, University of California Davis, Davis CA, 1981.

[19] W. P. M. van Swaaij, "Gasification: The process and the technology," *Resources and Conservation*, vol. 7, pp. 337–349, 1981.

[20] K. Littlewood, "Gasification: Theory and application," *Progress in Energy and Combustion Science*, vol. 3, no. 1, pp. 35–71, 1977.

[21] T. B. Reed, *A survey of biomass gasification*, vol. 3, Department of Energy, Solar Energy Research Institute, Golden, CO, 1979.

[22] M. Wood and P. Branch, *Wood gas as engine fuel*, Food and Agriculture Organization, 1986, Rome Italy.

[23] Z. Liu, "Gasification of municipal solid wastes: A review on the tar yields," *Energy Sources, Part A: Recovery, Utilization, and Environmental Effects*, vol. 41, no. 11, pp. 1296–1304, 2019.

[24] J. Venselaar, *Design rules for down-draught gasifiers, a short review*, IT Bandung, Indonesia, 1982.

[25] T. B. Reed, B. Levie and M. S. Graboski, *Fundamentals, development and scaleup of the air-oxygen stratified downdraft gasifier*, Solar Energy Research Institute, Golden, CO, 1987.

[26] J. H. W. Lee and V. H. Chu, *Turbulent jets and plumes: A lagrangian approach*, Kluwer Academic Publisher, Dordrecht, The Netherlands, 2003.

7 Trapped Vortex Combustor for Gas Turbine Application

Debi Prasad Mishra and PK Ezhil Kumar

CONTENTS

DOI: 10.1201/9781003049005-7

7.1   WHY A TRAPPED VORTEX COMBUSTOR?

In this section, we will answer the question, what is the need for a trapped vortex combustor? In order to address this question, first we have to briefly look at the basics of conventional swirl combustors and the difficulties associated with them. In a conventional swirl-stabilized combustor (see Figure 7.1a), the main flame is established in the primary zone with the aid of physical swirlers [Mishra (2015)]. As a result of continuous combustion, carbon monoxide (CO), carbon dioxide (CO_2), unburned hydrocarbons (UHC), particulate matters (PM) and oxides of nitrogen (NO_x) were emitted significantly [Lefebvre and Ballal (2010)]. Modern swirl combustors with improved designs could reduce these pollutant emissions to appreciable levels except NO_x [Lefebvre and Ballal (2010)]. A lean premixing concept was widely advocated as a useful strategy for NO_x reduction in gas turbine combustors [Muruganandam et al. (2004)]. While adapting this technology, the lean flame in the primary zone is exposed to mainstream flow and as a result, the flame is highly sensitive to the inlet conditions. Since the stability of the flame is limited by the mainstream condition, the operating range of these combustors will be narrow. This issue can be alleviated by using the trapped vortex combustor (TVC) concept put forward by Hsu et al. (1995, 1998). The schematic of a 2D trapped vortex combustor [Hendricks et al. (2001)] is shown in Figure 7.1b, where the mainstream air enters through the main inlet and a fraction of which is separated out at the cavity leading edge (see Figure 7.1b). Fluid, which is being separated at the cavity leading edge, is trapped inside the cavity as shown in Figure 7.1b. In order to sustain continuous combustion, fuel and air are injected directly into the cavity. Since the reactions occurring in the cavity are not exposed to mainstream flow, the pilot (cavity) flame is expected to be stable for a wide operating range as compared to that of a swirl-stabilized combustor. An attempt has been made to compare the significant differences between the conventional and TVC in the subsequent sub-section.

7.1.1   CONVENTIONAL VS. TRAPPED VORTEX COMBUSTOR

The main feature that differentiates the conventional combustor from a TVC is the method of flame stabilization. In order to bring out the significant differences between them, a photograph of the swirl flame from Muruganandam and Seitzman (2005) and the TVC flame from Ezhil (2014) are compared qualitatively in Figure 7.2a and b respectively. In the swirl combustor, the main flame is literally anchored at the downstream region of the swirler [Muruganandam and Seitzman (2005) and Tuncer et al. (2014)] (see Figure 7.2a), whereas the trapped vortex combustor has no physical holder to anchor the main flame; rather it is being continuously ignited by the cavity/pilot flames in as shown in Figure 7.2b.

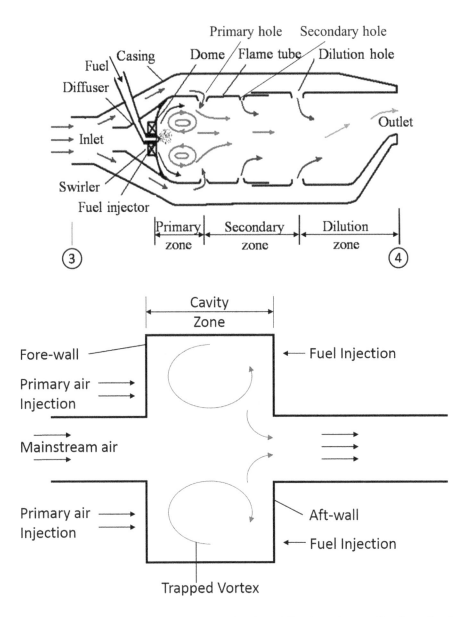

FIGURE 7.1 (a) Schematic of a conventional swirl stabilized combustor; (b) schematic of trapped vortex combustor (TVC) and the associated flow structure.

Source: Author's drawing.

Based on the observations from the flame photograph, the schematic of the swirl combustor flame and TVC flame are shown in Figure 7.2c and 7.2d respectively. It can be observed from Figure 7.2c that flow recirculation occurs downstream from the swirler [Driscoll and Jacob (2011), Kim et al. (2013) and Tuncer

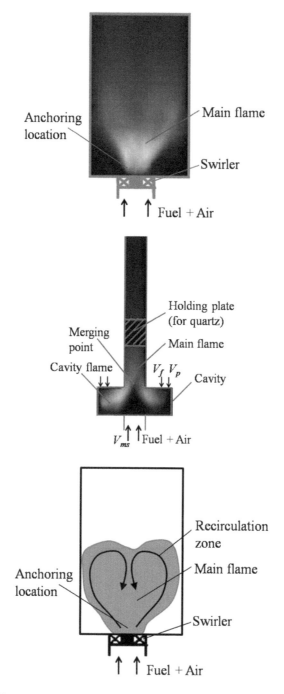

FIGURE 7.2 Illustration of significant differences between the stabilization mechanism of a swirl flame and a TVC flame; (a) swirl stabilized flame; (b) photograph of pilot ignited TVC flame; (c) swirl stabilized flame; (d) cavity/pilot stabilized TVC flame.

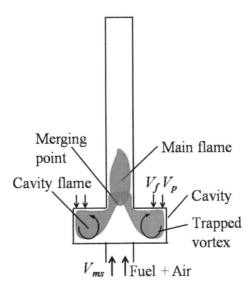

Merging point

Main flame

Cavity flame

$V_f V_p$

Cavity

Trapped vortex

V_{ms} ↑ ↑ Fuel + Air

FIGURE 7.2   (Continued)

et al. (2014)], where the local flow velocity is likely to be of the order of local burning velocity of the flame and hence this zone is favorable for flame stabilization. On the other hand, in TVC, a recirculation zone is established within the cavity, where the pilot flame is stabilized as shown in Figure 7.2d. This pilot flame will eventually ignite the incoming mainstream fuel-air mixture under varied operating conditions. From a practical point of view, the main attractive feature of the TVC is its superior performance characteristics specifically in terms of combustion efficiency, LBO limit and emission level as compared to the existing swirl combustors (see Table 7.1). For example, 99% of combustion efficiency can be achieved in TVC as compared to 95% in conventional combustors for an overall equivalence ratio of around 0.3 (see Table 7.1). Similarly, the emission level of the TVC is very low (25g/kg of Jet-A fuel) as compared to the swirl stabilized combustor (250 g/kg of fuel) for the same primary zone equivalence ratio [Roquemore et al. (2001)]. From Roquemore et al. (2001), for a Longwell loading parameter of 200, the LBO equivalence ratio limit of a conventional combustor is around 0.8 and for TVC, the LBO limit is approximately 0.3. This brief comparison indicates that the TVC offers more advantages than the conventional combustors from both design and performance perspectives.

Owing to the potential advantages offered by the trapped vortex combustor, researchers in this field have attempted to (i) comprehend the underlying process governing this combustor and (ii) explore the ways and means of enhancing its performance. Besides this, some research groups are attempting to modify of the existing design to augment the performance characteristics of TVC. However, identifying the crucial areas where more understanding is still required can only be brought out by carrying out a comprehensive review of available literature.

TABLE 7.1
Comparisons between Conventional and Trapped Vortex Combustor.

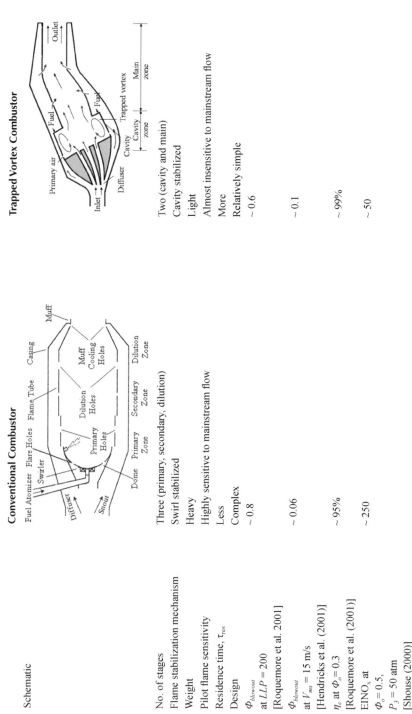

	Conventional Combustor	Trapped Vortex Combustor
Schematic		
No. of stages	Three (primary, secondary, dilution)	Two (cavity and main)
Flame stabilization mechanism	Swirl stabilized	Cavity stabilized
Weight	Heavy	Light
Pilot flame sensitivity	Highly sensitive to mainstream flow	Almost insensitive to mainstream flow
Residence time, τ_{res}	Less	More
Design	Complex	Relatively simple
$\Phi_{blowout}$ at $LLP = 200$ [Roquemore et al. 2001]	~ 0.8	~ 0.6
$\Phi_{blowout}$ at $V_{ms} = 15$ m/s [Hendricks et al. (2001)]	~ 0.06	~ 0.1
η_c at $\Phi_o = 0.3$ [Roquemore et al. (2001)]	~95%	~99%
EINO$_x$ at $\Phi_o = 0.5$, $P_3 = 50$ atm [Shouse (2000)]	~250	~50

Note: LLP = Longwell loading parameter $\Phi_{blowout}$ is the equivalence ratio at flame blowout, V_{ms} is the mainstream velocity, Φ_o is the overall equivalence ratio, η_c is the combustion efficiency and P_3 is the static pressure at combustor inlet.

Hence, in this chapter, a critical analysis of various aspects of the trapped vortex combustor such as the cavity flow structure, the fuel-air mixing, combustion efficiency, flame blowout limit, etc., under various operating as well as geometric conditions are explored. Furthermore, from the results reported in the literature, an attempt is being made to bring out the positive developments as well as lacunas to be addressed in the near future to bring this technology to a matured level. Before venturing into the technical studies on TVC, it will be interesting to look at a brief history on the development of TVC as enumerated in the subsequent section.

7.2   BRIEF HISTORY OF TRAPPED VORTEX COMBUSTOR

Huellmantel et al., a long time back in 1957, demonstrated the use of a cavity in stabilizing a premixed flame under subsonic conditions [Hsu et al. (1998)]. Using this cavity stabilized flame, they could improve the stability margin and reduce pressure drop across the combustor. In another independent study, Little and Whipkey (1978) investigated the usage of cavity in reducing the drag under non-reacting flow condition. They brought out the effects of cavity size on the drag and showed that trapping a stable vortex in the cavity was the cause for lower drag value. They also proposed certain sizing rules for trapping a stable vortex within the cavity. Based on these inputs, Hsu and co-workers (1995, 1998) made initial attempts to develop an axisymmetric (can type), internal cavity trapped vortex combustor. Roquemore et al. (2001) termed this variant as first generation TVC. Subsequently, the second (axisymmetric, external cavity) and third generation (2D opposed twin cavity) trapped vortex combustors had evolved as discussed in the following sub-sections.

7.2.1   FIRST GENERATION TVC

Motivated by the studies mentioned earlier, Hsu and coworkers [Hsu et al. (1995, 1998)] developed an axisymmetric tubular type trapped vortex combustor, in which the internal cavity was being employed as shown in Figure 7.3. According to Hsu et al. (1998), the major difference between Huellmantel et al. (1957) and their combustors was that in the former work, fuel and air were not injected into the cavity. In the latter work, fuel and air were injected into the cavity [Hsu et al. (1998)]. In TVC literature, the combustor developed by Hsu and co-workers is termed as the first generation (internal cavity) TVC [Roquemore et al. (2001)]. This first generation TVC exhibits lower pressure drop, better stability limit and lower emission level as compared to the conventional swirl stabilized combustor. Further, in this configuration, Stone and Menon (2000) carried out a numerical study to understand the fuel-air mixing in this axisymmetric TVC. Subsequently, Ezhil and Mishra (2012) studied the effects of momentum flux ratio and heat addition on the flow structure of this first generation TVC. The major issues with this combustor are (i) the obstruction of mainstream due to the presence of the fore-body and (ii) heating up of central assembly due to the presence of flame.

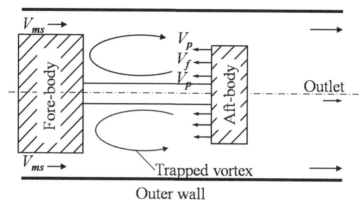

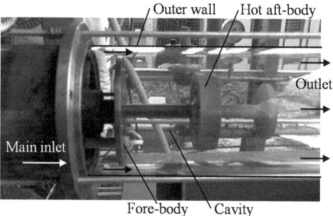

FIGURE 7.3 Schematic of first generation trapped vortex combustor.

7.2.2 SECOND GENERATION *TVC*

In order to overcome the blockage issues with internal cavity TVC, a can type combustor with a cavity oriented outside the main burner was designed and developed by a group at Air Force Research Laboratory, AFRL [Roquemore et al. (2001)] as shown in Figure 7.4a and 7.4b. This kind of combustor is called a second generation (external cavity) TVC [Roquemore et al. (2001)]. Much information about this combustor is not available in open literature except the scanty description provided by Roquemore et al. (2001). They also reported that this second generation TVC had better flame stability characteristics, lower pressure drop and operated efficiently over a wider operating range than the conventional combustor at atmospheric pressure.

Recently, this second generation can type TVC has been relooked by Gutmark and coworkers [Gutmark et al. (2007)]. However, in place of the diffuser, they have used a bluff-body for stabilizing the flame. In this kind of second generation TVC,

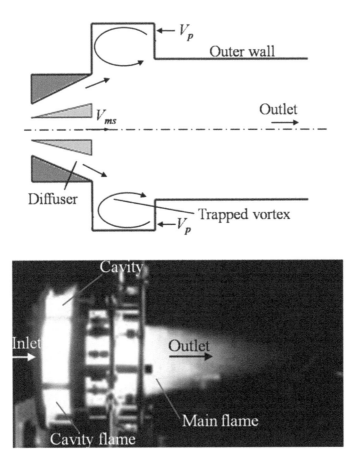

FIGURE 7.4 (a) Schematic of second generation trapped vortex combustor and (b) direct flame photograph from second generation TVC flame.

Source: Hendricks et al. (2005).

the flow in the circumferential cavity either oriented internally or externally is three dimensional in nature and fuel/air injection into the cavity will further complicate the situation, which will make it difficult for line-of-sight visualization and analyzing the flame for developing better understanding. Besides this it will be difficult to experiment with different injection strategies with this configuration. In order to facilitate these needs, the third generation TVC has been devised as discussed in the following paragraph.

7.2.3 THIRD GENERATION TVC

The third generation TVC is a 2D combustor configure sector (see Figure 7.5a and 7.5b), where different injection strategies can be investigated. Besides this, flow and flame visualization can be carried out easily for better understanding. For

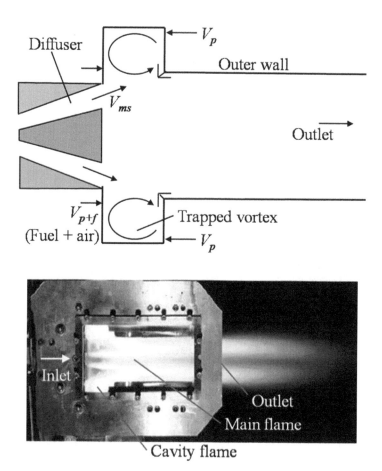

FIGURE 7.5 (a) Schematic of third generation trapped vortex combustor and (b) direct flame photograph of third generation TVC flame.

Source: Hendricks et al. (2005).

this combustor, Roquemore et al. (2001) mentioned that 19 injection configurations had been studied so far; however, the details of these configurations are not revealed in open literature. Among these, the best three injection configurations are discussed by Roquemore et al. (2001). They also reported that the third generation TVC performed better than conventional swirl combustors while operating at the same conditions. Later, similar combustor design was used by Hendricks et al. (2001), Atul and Ravikrishna (2011) and Ezhil and Mishra (2016, 2017a, 2017b) to study the various aspects of TVC under reacting and non-reacting flow conditions.

In this section, although an outline of various stages of development of TVC is discussed briefly, understanding of various characteristics of this combustor are crucial in its development and are discussed in detail in the subsequent sections.

7.3   CAVITY SIZE STUDIES

The very first step in the design of a TVC is to arrive at the cavity size. In this chapter, we will answer the following question, how to arrive at the best possible cavity size?

Cavity size is characterized by cavity aspect ratio (AR), which is defined as the ratio between cavity length (L) to cavity depth (D) (see Figure 7.1b). Katta and Roquemore (1998a) carried out a numerical study by considering three cavity aspect ratios (AR)—namely, $AR = 0.7$, 1.3 and 1.8. For all cases, the cavity injection location, injection velocity and mainstream velocity are maintained constant. For the $AR = 0.7$ case, multiple vortices were observed within the cavity. On the other hand, the $AR = 1.8$ case resulted in a single vortex within the cavity; however, this vortex was shifting its location with time [Katta and Roquemore (1998a, 1998b)]. They noted a single stream-wise vortex (*vortex rotating in the direction of mainstream flow*) was established for the $AR = 1.3$ case. In this case, the influence of mainstream flow on the trapped vortex is minimum as compared to other AR cases. Further, the pressure drop for this case was also minimum [Hsu et al. (1998) and Ezhil and Mishra (2012)].

Mishra and Sudharshan (2009) studied numerically the effects of four aspect ratios—namely, 0.9, 1.0, 1.2 and 1.4—on the non-reacting cavity flow structure of a 2D TVC. For $AR \leq 1.0$ cases, the cavity flow is characterized by multiple vortices. Similarly, multiple vortices are observed for the $AR = 1.4$ case as well. They observed a single dominant vortex of the size of the cavity, termed as a primary vortex, along with small secondary corner vortices formed for $AR = 1.0$ and 1.2 cases. They also observed that this primary vortex was stable both spatially and temporally. Their result complies with the result of Katta and Roquemore (1998a, 1998b) and Hsu et al. (1998). Mishra and Sudharshan (2009), from their numerical simulation, reported that stable reactions happened in the $AR = 1.2$ cavity.

Xing et al. (2012) compared the effects of three aspect ratios—namely, $AR = 1.14$, 1.25 and 1.43—on cavity flow structure of annular single cavity TVC. Through numerical simulations, they observed a single vortex for all three aspect ratio cases. However, the position of vortex core for $AR = 1.14$ and 1.43 cases got changed with time and hence stability is of concern for these two cases [Xing et al. (2012)]. On the other hand, vortex core for the $AR = 1.25$ case is located at the mid of the cavity over a certain period of time. Hence, they argued that the $AR = 1.25$ case is the optimum cavity size as compared to the other two cases. Ezhil and Mishra (2018) conducted a numerical study for the $AR = 1.2$ cavity and under various mainstream Reynolds number (Re_{ms}) conditions, the trapped vortex was stable and mainstream entrainment was minimum.

From this discussion, it is clear that to establish a stable flow inside the cavity, aspect ratio should be maintained from 1.0 to 1.3. Aspect ratios lesser than 1 are likely to result in shorter lead time, which eventually pump out fuel and leave the cavity fuel deficient. Now, the question is, will cavity mass addition disturb the flow structure? If so, what is the better way to inject fuel and air into the cavity? Hence the effects of injection strategies on the cavity flow structure are discussed in the following section.

7.4   UNDERSTANDING NON-REACTING FLOW CHARACTERISTICS

It is important to study the non-reacting characteristics of the cavity flow for various geometric and input conditions. This understanding will be helpful to interpret the characteristics under reacting flow conditions. Hence in the following sub-sections, various non-reacting flow characteristics are reviewed.

7.4.1   FLOW STRUCTURE

Several injection strategies are being used to establish a trapped vortex in TVC cavity [Hsu et al. (1998), Katta and Roquemore (1998a), Roquemore et al. (2001) and Mishra and Sudharshan (2009)]. It can be noted that the trapped vortex established within the passive cavity can be strengthened further by injecting air jet into the cavity. Alternatively, the direction of the cavity flow can be reversed by suitably placing the high velocity cavity air jet [Meyer et al (2002), Bucher et al. (2003), Straub et al. (2005), Wu et al. (2009) and Atul and Ravikrishna (2011)]. Based on the direction of trapped vortex, the cavity flow can be broadly classified into three types—namely, (i) stream-wise vortex, (ii) counter stream-wise vortex and (iii) multiple vortices.

7.4.2   ENTRAINMENT STUDIES

The relative momentum flux between the cavity and the mainstream flow affects entrainment in a trapped vortex combustor. The cavity flow might possess zero, positive or negative momentum; accordingly, the entrainment will be neutral (no entrainment), negative (out of the cavity) or positive (into the cavity) respectively [Sturgess and Hsu (1997)]. Besides this, the cavity entrainment will also be dependent on the adopted injection scheme, injector location as well as the momentum flux between cavity and mainstream flow. The mainstream air that entrains into the cavity may alter the local equivalence ratio within the cavity. Moreover, the extent of fuel-air mixing, flame stability, temperature distribution, combustion efficiency, emission, etc. can be affected by the amount of air entrained into the cavity. Hence, in this section, emphasis is being given to review the works on the entrainment of mainstream air into the cavity in a trapped vortex combustor.

7.4.2.1   Cavity Size Effects

In order to quantify the amount of mainstream air entrainment into the cavity experimentally, Sturgess and Hsu (1997) used flame blowout data. They injected fuel and air into the cavity and assumed that at the instant of flame blowout, all reactions are confined within the cavity region. If the fuel-air ratio at the instant of flame blowout is equal to the lean flammability limit of the mixture, then there is no entrainment into the cavity. On the other hand, if the blowout occurred early, then they argued that mainstream air must be entrained into the cavity. It can be noted that the flame may not be confined into the cavity for all the cases. More importantly, flame extinction within the cavity region may also occur due to the excessive strain rate, which is not being considered by these authors. However, in this section, the results reported in Sturgess and Hsu (1997) are discussed.

For low aspect ratio cases, they observed negative entrainment (outflow from cavity to the mainstream); the cause might be due to the paucity in cavity to accommodate even the injected fuel and air. As a result, the residence time would be less. For higher aspect ratio cases, maximum entrainment occurred [Sturgess and Hsu (1997)], which is confirmed by the numerical studies of [Katta and Roquemore (1998a)]. From this numerical work, it was also revealed that minimum amount of entrainment took place between the cavity and mainstream for optimum aspect ratio, AR of 1.3. Hence the mainstream entrainment for this case will not affect the cavity flow structure.

7.4.2.2   Mainstream and Primary Air Effects

Sturgess and Hsu (1997) also studied the effects of mainstream and primary air velocity (V_p) on the entrainment in an axisymmetric trapped vortex combustor. For $V_p = 0$ m/s case (passive cavity), the entrainment into the cavity increased with increase in mainstream air [Sturgess and Hsu (1997)]. On the other hand, for the $V_p = 42$ m/s case (active cavity), an increase in mainstream velocity reduced the cavity entrainment. For the case where cavity and mainstream air velocities were comparable, the entrainment was found to be lower. They argued that the relative momentum flux between the cavity and the mainstream air is the main factor that controls entrainment. Besides this, for a particular mainstream velocity, increasing the primary air velocity initially lowered down the cavity entrainment until a critical V_p value. Subsequent increase in primary air velocity augmented the cavity air entrainment. The cause for this observation was not investigated adequately by the authors.

7.4.3   CAVITY JET PENETRATION TO MAINSTREAM

The cavity flame can act as an ignition source for the mainstream fuel-air mixture. It can be noted that if the cavity flow is not having sufficient momentum, then the mainstream flow would quench the cavity flame, which is trying to ignite the mainstream fuel-air mixture. Hence the cavity flow should have sufficient momentum to penetrate into the mainstream zone. To understand further, the effects of various operating conditions on the penetration depth for a stream-wise injection case under non-reacting flow conditions are considered in this section.

The extent of uniformity in the turbulence level at the downstream locations (relative to the cavity trailing edge) is sensitive to the penetration depth of the cavity fluid into the mainstream region. Figure 7.6 shows the trajectories of the massless particles injected from the cavity. It can be noted from Figure 7.8 that a fraction of the injected particles recirculates into the cavity and the remaining particles penetrate into the mainstream region. Generally, penetration depth is defined as maximum distance by which the cavity fluid penetrates into the mainstream flow. The penetration depth increases with increase in primary air-flow rate as mentioned in Table 7.2. Hence, it is expected that the mixing of cavity fluid with the mainstream flow can be enhanced by increasing the primary air jet velocity.

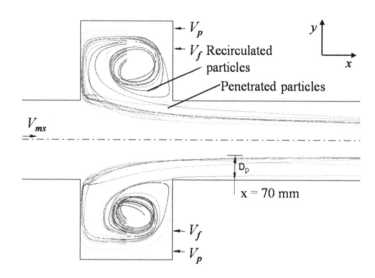

FIGURE 7.6   Particle trajectories from the cavity for the $Re_{ms} = 20000$, $V_p = 40$ m/s case.

TABLE 7.2
Effect of Cavity Air Velocity on Penetration Depth, D_p.

Case No.	Re_{ms}	V_p	Penetration Depth (D_p), mm
5	20000	40	5.9
6	20000	60	8.3
7	40000	60	5.5
8	40000	100	7.7

7.4.4   FUEL-AIR MIXING

Mixing of different streams carrying fuel and air is of practical interest, especially for a trapped vortex combustor, since the chemical reactions in a cavity are basically mixing limited. As a result, the overall performance of a trapped vortex combustor is dictated by the extent of molecular mixing of fuel and air within the cavity. The effectiveness of mixing can be described by using the parameter intensity of segregation, I_s (see Eq. 1).

$$I_s = \frac{\overline{Y_f'^2}}{\overline{Y_f}\,\overline{Y_{Ox}}} \tag{1}$$

where $\overline{Y_f'^2}$ is the time rate of change of fuel mass fraction, $\overline{Y_f}$ is the mean fuel mass fraction and $\overline{Y_{Ox}}$ is the mean oxidizer mass fraction. This method cannot be employed to the cases where macro mixing is absent [Danckwerts (1958) and Mishra and Sudharshan (2009)]. Hence, sufficient care has to be taken while ascertaining the extent of mixing using this method.

7.4.4.1 Cavity Sizing Effects on Mixing

Mishra and Sudharshan (2009) studied the effects of cavity aspect ratio on the mixing quality by considering three aspect ratios—namely, AR = 1.0, 1.2 and 1.4. Their study revealed that a cavity AR of 1.2 resulted in lower intensity of segregation, indicating better fuel-air mixing. In contrast, non-optimum AR cases resulted in higher I_s values indicating the deterioration of mixing, which is caused due to the presence of multiple vortices.

7.4.4.2 Mainstream Velocity Effects on Mixing

Stone and Menon (2000) computed the variation of rms fuel mass fraction as a function of mainstream velocity to study the extent of fuel-air mixing in the TVC cavity. They reported higher level of fluctuations in the fuel mass fraction for low mainstream velocity (V_{ms} = 20 m/s) and the magnitude of the fluctuating quantity reduced while increasing the mainstream velocity to about 40 m/s. They argued that for higher mainstream velocities, the fuel-air mixing rate increased, leading to a reduction in mixing time.

These two studies reveal the influence of cavity size and main inlet velocity on the fuel-air mixing within the cavity. However, the mixing between the cavity stream and the mainstream flow is very important and is not being addressed in these works.

7.4.4.3 Cavity Flow Scheme Effects on Mixing

The parameter called spatial mixedness (ζ) is used to quantify the mixing at a particular plane. The parameter ζ is obtained by minor modification of the intensity of segregation and is defined as

$$\text{Spatial mixedness, } \zeta = 1 - \left(\frac{\sigma_f^2}{\mu_f \left(1 - \mu_f \right)} \right) \tag{2}$$

where σ_f and α_f are the standard deviation and mean of the fuel mass fraction respectively at the plane of interest. It can be noted that the value of ζ can vary between 0 and 1. The mixedness parameter, ζ = 0, corresponds to unmixed fuel and air streams and ζ = 1 will indicate uniform mixing. This method cannot be employed at the fuel and air inlets where it does not have any physical meaning.

Differences in spatial mixedness (ζ) at the downstream location for two injection schemes—namely, (i) stream-wise (SW) injection case and (ii) counter stream-wise (CSW) injection case at a mainstream Reynolds number of 20000—are compared in Figure 7.7. It can be observed that at downstream locations of the cavity trailing edge, the mixing index value for the CSW case is higher than that of the SW case. Most of the cavity fuel for the CSW case moves out from the cavity and mixes with the mainstream flow (see Figure 7.8b), leaving a major portion of the cavity region fuel deficient.

On the other hand, for the SW case, most of the injected fuel is retained within the cavity and a lesser amount of fuel gets transported into the mainstream region as shown in Figure 7.8a. This analysis provides a notion that for the *CSW* case the cavity is observed to be fuel deficient besides its non-uniform fuel-air distribution. Because of the fuel deficient cavity zone, the stability margin for this case is likely to be very narrow.

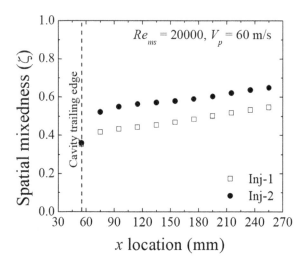

FIGURE 7.7 Effects of injection scheme (Inj-1: stream-wise; Inj-2: counter stream-wise) on the spatial mixedness at different planes along the stream wise direction.

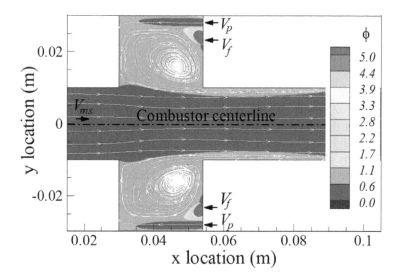

FIGURE 7.8 Equivalence ratio contours with streamlines (white color) embedded on it for (a and c) stream-wise injection case and (b and d) counter stream-wise injection case; (a) $Re_{ms} = 20000$, $V_p = 40$ m/s; (b) $Re_{ms} = 20000$, $V_p = 40$ m/s; (c) $Re_{ms} = 20000$, $V_p = 60$ m/s; (d) $Re_{ms} = 20000$, $V_p = 60$ m/s.

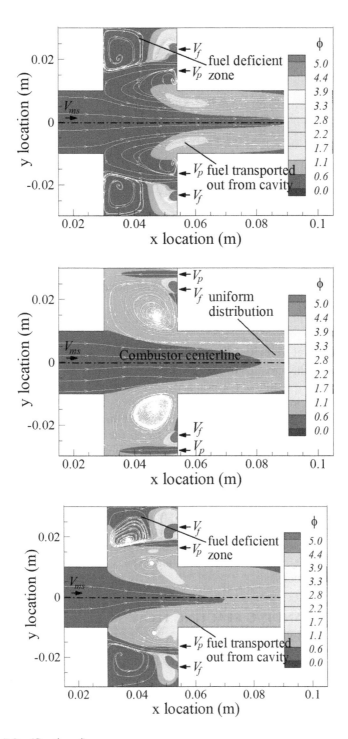

FIGURE 7.8   (Continued)

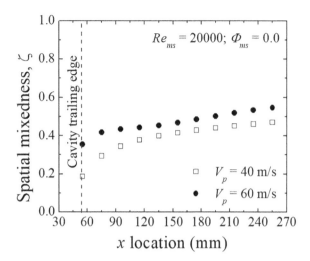

FIGURE 7.9   Variation of spatial mixedness at different planes along the stream-wise direction for the *SW* case.

7.4.4.4   Cavity Jet Velocity Effects on Mixing

Figure 7.9 shows the variation of spatial mixedness (ζ) along various x locations for the *SW* case at Re_{ms} = 20000. For primary air velocity of 40 m/s, ζ value turns out to be 0.2 at a location just downstream from the cavity trailing edge and increases along the combustor axis as shown in Figure 7.9. At the downstream locations, the fuel and air stream mixes with each other due to the residual momentum fluxes between the cavity and mainstream flow. However, enhanced mixing is very essential at the vicinity of the cavity trailing edge. When the primary air jet velocity is increased to 60 m/s, the mixing between the two streams is enhanced at the vicinity of the cavity trailing edge as well as at the downstream locations. From the previous analysis, it is very clear that the cavity to mainstream momentum flux plays a major role in the cavity-mainstream interactions. Therefore, momentum flux will also have a definitive role in combustion characteristics—namely, the flame length, flame stability limit, combustion efficiency and exhaust emissions—as discussed in the subsequent sections.

7.5   UNDERSTANDING REACTING FLOW CHARACTERISTICS

In the previous section, an attempt was made to understand the various aspects of trapped vortex combustors under non-reacting flow conditions. In this section, important reacting flow characteristics—namely, the flame length, flame stability limit, pollutant emission level and combustion noise—are discussed. At the end of this section, an interesting topic, the method of dynamic sensing of lean blowout for TVC, is also brought out.

7.5.1   FLAME LENGTH

Flame length is an important aspect to be considered in the design phase of gas turbine combustors. Generally, the downstream length of the combustor will be decided by the flame length. It can be noted that conventional combustors are 15 to 25 cm long [Hsu et al. (1998)] and hence it is expected that the flame length should be of this order. Hsu et al. (1998) investigated the effect of cavity fuel flow rate on the flame length in an axisymmetric trapped vortex combustor (TVC) as shown in Figure 7.10a. As the cavity fuel flow rate is increased for the V_{ms} = 14 m/s case, the

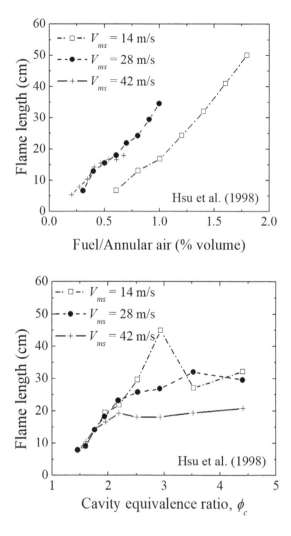

FIGURE 7.10 Variation of flame length with (a) cavity fuel and (b) cavity equivalence ratio (by varying cavity air).

Source: Hsu et al. (1998).

excess fuel is likely to escape from the cavity and burn in the downstream region of the combustor leading to the increased flame length. Besides this, the flame length was also sensitive to mainstream air velocity [Hsu et al. (1998)].

According to Hsu et al. (1998), at higher mainstream air, a larger amount of cavity fuel was transported to the mainstream air and was ignited downstream. However, the effect of cold mainstream air on flame quenching was not discussed by the authors. The flame length was observed to increase in the range of cavity equivalence ratio ($\Phi_c = 0.5$ to 2.2). Beyond this range, flame length variation with Φ_c is insignificant as shown in Figure 7.10b. Though the authors studied the effects of various operating conditions on the flame length, the effect of mainstream premixing is not considered. This earlier discussion revealed that flame length is sensitive to cavity equivalence ratio, mainstream velocity and cavity fuel flow rate. However, primary air velocity is also likely to influence the flame length in a trapped vortex combustor [Ezhil and Mishra (2017a)]. Figure 7.11 shows the variation of flame length with primary air jet velocity for different mainstream Reynolds number cases. For a particular cavity equivalence ratio, the flame length increases with primary air velocity as shown by Figure 7.11. In order to maintain the cavity equivalence ratio constant, the cavity fuel flow rate has to be increased proportionally with primary air velocity leading to an increase in overall power level and hence, the flame length tends to increase. Besides this, an increase in primary air velocity also augments the burning of mainstream fuel-air mixture, which is also instrumental in enhancing the flame length value.

7.5.2   LEAN FLAME STABILITY LIMIT

The flame stability loop delineates the stable and unstable operating regimes of the combustor. Flame blowout limit or lean flame stability limit is the minimum fuel flow rate below which the flame will be blown out of the combustor [Mishra (2010) and Lefebvre and Ballal (2010)]. Flame stability limit of a gas turbine combustor is influenced by a number of parameters such as the pressure, temperature, combustor geometry, equivalence ratio and the free stream velocity.

7.5.2.1   Effects of Primary Air Velocity

Hsu et al. (1998) studied the effects of cavity air-flow rate on the blowout limits of an axisymmetric trapped vortex combustor for two mainstream velocity cases—namely, (i) 7 and (ii) 42 m/s. In this study, the mainstream air-flow rate and cavity air-flow rate were fixed and the fuel flow rate was reduced till the occurrence of blowout. They observed that the blowout limit became wider for low mainstream velocity (7 m/s) as compared to the higher mainstream velocity case (42 m/s) as shown in Figure 7.12.

It can be noted that the stability curve exhibits a *concave upward* trend for both the extreme mainstream velocity cases. At a primary air velocity of 25m/s, both curves merge together. Unfortunately, the cause for this interesting trend was unexplained by those authors. However, an attempt was made by the present authors

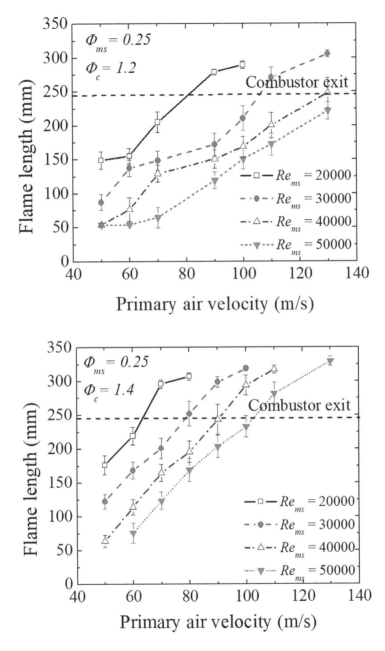

FIGURE 7.11 Effects of primary air velocity on flame length for two cavity equivalence ratios—namely, (a) $\Phi_c = 1.2$ and (b) $\Phi_c = 1.4$.

Source: Ezhil and Mishra (2017a).

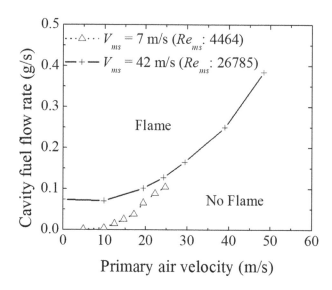

FIGURE 7.12 Effect of primary air velocity on the static flame stability limit of an axisymmetric trapped vortex combustor.

Source: Hsu et al. (1998).

to interpret this trend. According to our experience, one of the dominant factors influencing the stability margin is the 'momentum flux ratio' between the cavity and mainstream flow [see Eq. 3]. In this case, until primary air velocity 26m/s, the mainstream momentum may be dominant, and beyond primary velocity 26m/s, the cavity injection momentum dominates the mainstream. It can also be noted that injection strategy adopted by Hsu et al. (1998) resulted in a stream-wise vortex within the cavity and the blowout limit curve might be different for the counter stream-wise injection case and is not being explored.

Ezhil and Mishra (2016) also reported flame stability limit of a 2D TVC for a stream-wise flow case (see Figure 7.13). For a particular mainstream Reynolds number, the amount of fuel required to stabilize the flame increases with the primary (cavity) air velocity. The flame blowout trend reported in Hsu et al. (1998) and Ezhil and Mishra (2016) are similar in nature. It is interesting to note from Figure 7.13 that the flame blowout limit curve transforms from a concave downward trend at $Re_{ms} = 20000$ to a concave upward trend at $Re_{ms} = 50000$.

7.5.2.2 Influence of Cavity Size on Flame Blowout Limit

Hsu et al. (1998) also studied the effect of cavity aspect ratio on the flame stability limit. The blowout curve remained almost invariant for any change in the cavity length till the aspect ratio of 1.3, beyond which the blowout equivalence ratio increased drastically as shown in Figure 7.14. Xing et al. (2010, 2012) investigated the effect of cavity size on the flame blowout limit of the TVC cavity for two cavity aspect ratio (AR) cases—namely, $AR = 1.13$ and $AR = 1.43$.

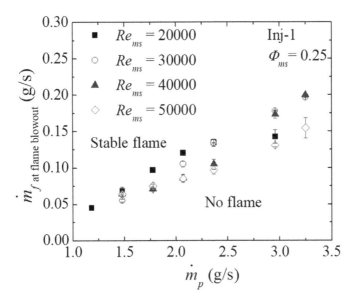

FIGURE 7.13 Effect of primary air mass flow rate on the flame blowout limits for stream-wise (Inj-1) case at mainstream equivalence ratio of 0.25.

Source: Ezhil and Mishra (2016).

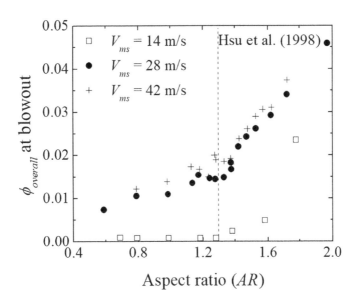

FIGURE 7.14 Effect of cavity aspect ratio on the static flame stability limit of an axisymmetric trapped vortex combustor.

Source: Hsu et al. (1998).

Based on the information available in Xing et al. (2012), we have calculated the cavity volume and related to their flame blowout limit data. For the $AR = 1.13$ case, a cavity with larger volume ($V = 6.05 \times 10^{-4}$ m^3) resulted in a much wider flame stability limit as compared to a cavity with smaller volume ($V = 4.9 \times 10^{-4}$ m^3). In contrast, for the $AR = 1.43$ case, a cavity with a smaller volume resulted in a wider stability limit as compared to the cavity with bigger volume as shown in Figure 7.13. The cause for this observation remains an open challenge. Wu et al. (2009) studied the effects of cavity air-flow rate on blowout limit of a hydrogen fueled 2D trapped vortex combustor. It was revealed that beyond a critical mainstream velocity, the blowout limit curve (cavity air-flow rate vs. cavity fuel flow rate curve) overlapped each other. From this trend, Wu et al. (2009) argued that beyond a critical mainstream velocity, the blowout limit would be influenced only by the cavity air and fuel flow rates and not by the mainstream air velocity.

It could be noted that in all these previously mentioned studies, the mainstream was not premixed with fuel. However, in gas turbine applications, the mainstream air is usually premixed with fuel. In order to assess the static flame stability limit of a single cavity trapped vortex combustor, Atul and Ravikrishna (2011) conducted an experimental study under varied operating conditions. They reported few data on flame blowout; however, the stability limit curve was not brought out from the study.

7.5.2.3   Effects of Momentum Flux Ratio

From the studies of Hsu et al. (1998) and Ezhil and Mishra (2016) in Section 7.5.2.2, it was revealed that the flame blowout limit is sensitive to primary air velocity. In order to understand this further, Ezhil and Mishra (2016) plotted the momentum flux ratio flame vs. lean flame blowout limit curve as shown in Figure 7.15. Momentum flux ratio (MFR) is defined as the ratio of momentum flux between the cavity and the mainstream flow and is given by

$$MFR = \frac{\left(\rho V^2\right)_p + \left(\rho V^2\right)_f}{\left(\rho V^2\right)_{ms}} \tag{3}$$

where V_p, V_f and V_{ms} are respectively the primary, cavity fuel and mainstream air velocities. Figure 7.17 shows the variation of cavity equivalence ratio with respect to the momentum flux ratio (MFR) at the instant of blowout.

For the mainstream equivalence ratio of the 0.25 case, at $Re_{ms} = 20000$, the blowout equivalence ratio increases till a critical MFR value and then it drops down as shown in Figure 7.15. This critical MFR value delineates two different physical situations— namely, the non-merged and merged cavity flames as shown in Figure 7.16.

$V_p = 40$ m/s $V_p = 60$ m/s $V_p = 80$ m/s $V_p = 100$ m/s

As the MFR reaches a critical value of ~45 (corresponding to $Re_{ms} = 20000$, $V_p = 70$ m/s case), both the cavity flames tend to merge together (see Figure 7.16). As a result, this merged flame is likely to ignite the mainstream fuel-air mixture. The

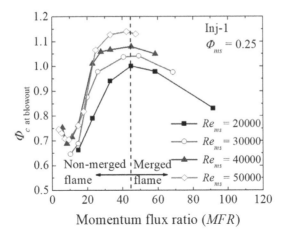

FIGURE 7.15 Effect of momentum flux ratio on the blowout limits for the stream-wise (Inj-1) case. *Source:* Ezhil and Mishra (2016a).

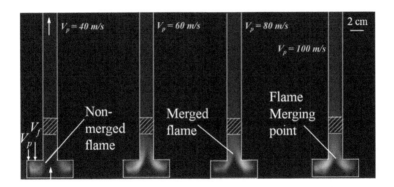

FIGURE 7.16 Flame images at the instant of blowout for the Re_{ms} = 20000 case.

mainstream flame thus established may widen the flame stability limit to a certain extent. Ezhil and Mishra (2016) also developed a correlation for the lean flame stability limit in terms of Damkohler number (Da) and MFR.

$$FSP = \frac{\dot{m}_{f,c}}{\dot{m}_{f,ms}} = 0.04\left(MFR^{0.5} / Da_c^{0.4}\right). \qquad (4)$$

7.5.3 EMISSION CHARACTERISTICS

Combustion products emitted from a gas turbine combustor consist of pollutants—namely, (i) carbon monoxide, (ii) carbon dioxide, (iii) oxides of nitrogen, (iv) unburned hydrocarbon [Mishra (2015)]. The level of emission of these pollutants depends on the operating conditions and the type of combustor design. In the context of a trapped vortex combustor, considerable efforts have been made by various groups to study the emission levels for varied operating conditions. In this section, an attempt has been made to review the various works on the emission characteristics of the trapped vortex combustors.

7.5.3.1 Effects of Cavity Equivalence Ratio

In an axisymmetric trapped vortex combustor, Hsu et al. (1998) studied the effects of cavity equivalence ratio on the emission levels for non-premixed mainstream air. Figure 7.17a shows the variation of emission indices of UHC, CO and NO$_x$ with

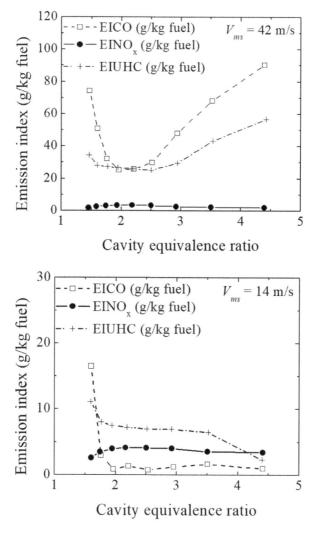

FIGURE 7.17 Variation of emission index with cavity equivalence ratio in single cavity axisymmetric TVC for two mainstream velocities—namely, (a) 42 m/s and (b) 14 m/s.

Source: Hsu et al. (1998).

cavity equivalence ratio for a primary air velocity 42 m/s in a single cavity TVC. It can be noted that EICO value decreases and reaches a minimum value and subsequently increases to higher value with increase in equivalence ratio. Variation of EIUHC also exhibits similar trend. At higher equivalence ratio, the excess fuel may not have sufficient air to oxidize it, and thus giving rise to the CO and UHC emissions. Besides this, the trend exhibited by the CO and UHC emissions for primary air velocity of 14 m/s is different from the higher velocity cases (see Figure 7.17b). In this case, as the residence time of the mainstream flow is lower, the hot products from the cavity could interact with the mainstream flow and eventually result in almost complete combustion. The NO_x emission for all the mainstream velocity cases is well below 10 g/kg fuel. As the mainstream air is not premixed, the flame emanating from the cavity region may get quenched and eventually the temperature is likely to be dropped down leading to a lower NO_x emission level. However, it is expected to have different emission levels for the premixed mainstream case as compared to the non-premixed case.

The effect of cavity equivalence ratio on the NO_x and CO emission level was also studied by Gutmark et al. (2007) in a can type external cavity trapped vortex combustor. In contrast to the work of Hsu et al. (1998), the low emission window was in the range of cavity equivalence ratio 0.75 to 1.5. It can be noted that for both the cases, the cavity vortex is in the direction of the stream-wise direction (see Section 7.4.1) and both the cases are non-premixed. However, the cause for this contradicting behavior is not well understood.

In another study by Straub et al. (2005), in a RQL trapped vortex combustor, the CO emission level is much lower as compared to the Gutmark et al. (2007) data (see Table 7.3). For a cavity equivalence ratio of the 1.6 case, the CO emission reported in Straub et al (2005) is almost zero, whereas it is around 1000 ppm for Gutmark et al. (2007). However, the NO_x values reported in Gutmark et al. (2007) are very low as compared to Straub et al. (2005). A stream-wise vortex is established in the cavity for Gutmark et al. (2007) and a counter stream-wise vortex for the injection strategy adopted by Straub et al. (2005). Though it is claimed by Straub et al. (2005) that the combustor is operating in RQL mode, the NO_x emission generated is much higher than the results of Gutmark et al. (2007). The emission data from these works are compared in Table 7.3. The causes for these differences in the emission levels for both the cases need further investigation.

7.5.3.2 Mainstream Equivalence Ratio Effects

The effect of mainstream equivalence ratio on the NO_x emission level have been studied by Gutmark et al. (2007) and Edmonds et al. (2006, 2008). Figure 7.20 shows the combined plot of the two works and both the data are in good agreement with each other for the same condition. It can be observed that the NO_x value varies with the mainstream equivalence ratio and lower values of the NO_x emission are obtained for mainstream equivalence ratios in the range of 0.4 to 0.6. Besides this, the effect of cavity equivalence ratio was considered by Edmonds et al. (2008) and their study revealed that the NO_x emission could be brought down by lowering the cavity equivalence ratio as indicated by the arrow in Figure 7.18.

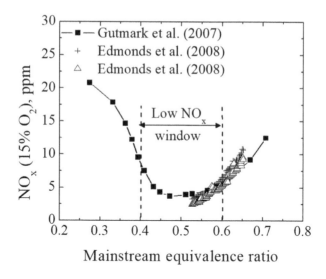

FIGURE 7.18 Effect of mainstream equivalence ratio on NO$_x$ emission level.

7.5.3.3 Effects of Residence Time

As mentioned earlier, residence time of the fuel-air mixture also affects the generation of the NO$_x$ emissions [Lefebvre and Ballal (2010) and Mishra (2010)]. In the
case of a trapped vortex combustor, the residence time of the flow in the cavity can
be changed by varying the cavity air-flow rate [Straub et al. (2005)] and almost in all
the mentioned studies, the residence time is varied in one form or other. However,
Hendricks et al. (2001) and Straub et al. (2005) explicitly showed the effect of cavity residence time on the NO$_x$ emissions (see Figure 7.19). In both cases, a counter
stream-wise vortex was established in the cavity.

Nonetheless, the NO$_x$ vs. residence time trend from both the work contradicts each
other (see Figure 7.19). It can be noted that the residence time in Hendricks et al. (2001)
is in the range of 3 to 9 ms while for Straub et al. (2005), it is in the range of 8 to 15 ms.
Hence it is difficult to arrive at a conclusion from these two studies. The summary of
the reviewed works are given in Table 7.3, and it can be noted that inconsistencies exist
in the units reported by various groups and hence a quantitative comparison may not be
possible with this information. This calls for a detailed systematic study on this subject.

7.5.4 ACOUSTIC CHARACTERISTICS

In the context of trapped vortex combustor, limited studies are reported for combustion noise; for example, Sturgess and Hsu (1997) through their experiments brought
out the difference in frequency spectrum between the reaction and non-reacting flow
noise. Apart from this study, not much data is available for combustion noise in a
trapped vortex combustor. Shivashankara and Crouch, a long time back in 1977,
brought out correlations between sound power and operating variables for a gas

TABLE 7.3
Emission Studies Carried Out by Various Researchers.

Sl No.	Author	Type of Combustor	Injection Pattern	Fuel	Operating Condition	Emission Range
1	Hsu et al. (1998)	Can type— Internal cavity	Stream-wise	Propane	V_{ms}: 14–42 m/s Φ_c: 1–4.5	EICO: 0–100 (g/kg fuel) EINO$_x$: 1.7–4.3 (g/kg fuel) EIUHC: 0–58 (g/kg fuel)
2	Straub et al. (2005)	Can type— External cavity	Counter stream-wise	Natural Gas	Φ_c: 1.2–2.6	CO: 0–10 (ppm) NO$_x$: 30–60 (ppm)
3	Gutmark et al. (2007)	Can type— External cavity	Stream-wise	–	Φ_c: 0.5–2.5	CO: 250–3100 (ppm) NO$_x$: 6–12 (ppm)
4	Hendricks et al. (2001)	2D—External cavity	Stream-wise	Jet-A	V_{ms}: 15–50 m/s F/A: 0.012–0.036	EINO$_x$: 2–11 (g/kg fuel)
5	Wu et al. (2009)	2D—External cavity	Counter stream-wise	Hydrogen	V_{ms}: 8.5–34 m/s Φ_c: 0.4–0.6	NO$_x$: 7–10 (ppm) (non-premixed) NO$_x$: 3–8 (ppm) (premixed)
6	Bucher et al. (2003)	2D—Internal cavity	Counter stream-wise	Methane	–	CO: 8–50 (ppm) NO$_x$: 5–60 (ppm)
7	Xing et al. (2010)	2D—External cavity	Stream-wise	Jet-A	M: 0.2 to 0.65 F/A: 0.0035–0.006	CO: 50–500 (ppm) NO$_x$: 2–20 (ppm)
8	Atul and Ravikrishna (2011)	2D—Single cavity	Counter stream-wise	Methane	V_{ms}: 8.4 m/s Φ_c: 0.8–1.3 Φ_o: 0.5	UHC: 0.8–1.9 (g/kg fuel) CO: 30.8–134 (g/kg fuel) NO$_x$: 0.01–0.5 (g/kg fuel)

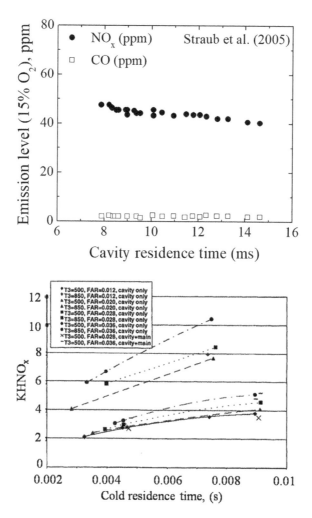

FIGURE 7.19 Effect of cavity residence time on (a) CO and NO$_x$ emissions; (b) NO$_x$ emissions.
Source: Straub et al. (2005); Hendricks et al. (2001).

turbine combustor. Similar correlations are not available for trapped vortex combustors and such correlations may be helpful to relate the influence of various parameters on the sound power. A new correlating parameter was devised by Ezhil and Mishra (2015) for the variation of *OASPL* with the correlating parameter as shown in Figure 7.20. They claimed that using this correlating parameter, *OASPL* curves for different operating conditions are collapsed into a single curve.

$$OASPL = 80 + 0.88\left(\frac{MFR^{0.5} \times \Phi_{overall}^{0.35}}{Da_c^{0.65}}\right) \qquad (5)$$

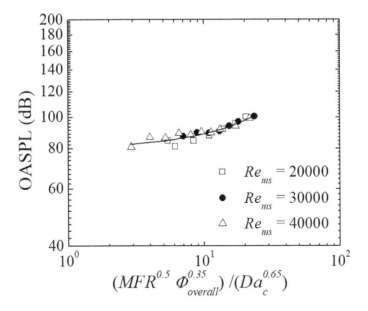

FIGURE 7.20   Correlation for overall sound pressure level (*OASPL*).

Source: Ezhil and Mishra (2015).

This correlation (Eq. 5) is valid for the range of operating conditions given: Re_{ms} = 20000 to 40000; V_p = 40 to 160 m/s; Φ_c = 1.0 to 1.4 and Φ_{ms} = 0.25.

7.5.5   LEAN BLOWOUT SENSING

In the context of swirl stabilized combustor, prior to flame blowout, unsteadiness of the flame emanating from the recirculation zone was reported by various groups [Muruganandam et al. (2005) and Yi and Gutmark (2007)]. This unsteadiness is nothing but a precursor event and can be used in the detection and control of flame blowout [Muruganandam et al. (2005) and Yi and Gutmark (2007)].

Since the TVC cavity flame is unstable prior to blowout, it will be interesting to look at the characteristics of the unstable flame near blowout. In this section, the near blowout studies carried out by Ezhil and Mishra are described. CH* chemiluminescence signatures from the cavity of a 2D trapped vortex combustor (TVC) are acquired and their spectral and statistical characteristics are analyzed by them. Frequency and time domain analyses of this signal revealed that at proximity to blowout, the combustor exhibits low frequency random fluctuations. Similar observations were reported in a lean premixed swirl stabilized combustor by Muruganandam et al. (2005). These fluctuations are found to be caused by repetitive extinction and re-ignition events. Moreover, the flame blowout phenomenon is purely random and no determinism exists at the conditions near blowout. In order to

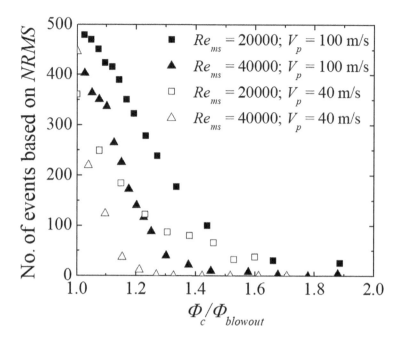

FIGURE 7.21   Number of precursor events based on the first threshold (*NRMS* < 0.25).

predict the approach of the flame blowout, event counts based on arbitrary threshold values are reported in literature [Muruganandam et al. (2005) and Yi and Gutmark (2007)]. This sensing methodology has certain limitations and hence a statistical quantity called the normalized root mean square (NRMS) value was employed by Ezhil and Mishra [see Figure 7.21]. This method of blowout sensing is considered to be quite robust and can be used as an active control method while operating TVCs under lean premixed conditions.

7.6   CONCLUDING REMARKS

The main emphasis of this chapter was to give the readers a common understanding about the fundamental aspects of trapped vortex combustors (TVC). First, an attempt was made to consolidate various early researches made by different groups so that the reader would appreciate the consistent efforts. Subsequently, the cavity size and the methodology of determination of cavity size was brought out through flow visualization and fuel-air mixing studies. It was brought out that cavity length to dept ratio in the range between 1.0 to 1.3 is most appropriate for TVC application. Followed by this, various injection schemes were analyzed and these injection schemes resulted in two flow patterns—namely, (i) stream-wise vortex flow and (ii) counter stream-wise vortex flow. Various methods of establishing both types of flow via cavity injection schemes are elaborately reviewed and the relevant merits and issues were also brought out. From the fuel-air mixing and flame stability point of view, reinforcing the naturally trapped vortex flow is preferred. Reinforcing the naturally formed vortex resulted in uniform fuel-air mixing as compared to that of the counter stream-wise flow. This indicated that stream-wise flow pattern would eventually result in a better stability margin, which is confirmed by the flame stability studies of Ezhil and Mishra (2016). The flame stability loop of a TVC was also influenced by the physical situation—namely, merged or non-merged flame. Review also brought out that the pollutant emission level from the TVC combustor is influenced by number of factors—namely, the power level, cavity/mainstream equivalence ratio and residence time. Flame merging will also likely have an influence on the emission level which needs further investigation. Near flame blowout detection studies highlighted that precursor events based on 'normalized root mean square' (NRMS) were suitable for active control implementation in TVC environment. In order to get best performance from TVCs, the following conditions—namely, (i) stream-wise vortex flow, (ii) cavity aspect ratio between 1 and 1.3, (iii) cavity momentum flux > mainstream momentum flux, (iv) merging cavity flame and (iv) fuel stream to be as close as the primary air jet—are to be ensured.

7.7 NOMENCLATURE

AC	autocorrelation coefficient
AR	aspect ratio
C_d	coefficient of discharge
C_P	specific heat capacity (J/kgK)
CI	confidence interval
CO	carbon monoxide
D	cavity depth (m)
Da	Damkohler number
D_c	cavity drag (N)
D_p	penetration depth (m)
d	diameter (m)
EDC	eddy dissipation concept model
EI	emission index (g/kg of fuel)
h	convective heat transfer coefficient (W/m^{2}K)
LES	large eddy simulation
I_{CH*}	$CH*$ chemiluminescence intensity
I_L	luminous intensity (from pixels)
I_s	intensity of segregation
I_t	turbulence intensity
k	turbulent kinetic energy (m^2/s^2)
L	cavity length (m)
L_C	combustor length (m)
L_f	flame length (m)
$\dot{m}$	mass flow rate (kg/s)
M	Mach number
MFR	momentum flux ratio

NO_x	oxides of nitrogen
NRMS	normalized root mean square
Nu	Nusselt number
OASPL	overall sound pressure level (dB)
P_t	total pressure (N/m²)
PDF	probability density function
PF	pattern factor
Pr	Prandtl number
PSD	power spectral density
Re	Reynolds number
RR	reaction rate
S_L	laminar burning velocity (cm/s)
T	temperature (K)
TVC	trapped vortex combustor
UHC	unburned hydrocarbon
$\bar{V}$	time mean x component of velocity (m/s)
$V_\circ$	free stream velocity (m/s)
V	instantaneous velocity (m/s)
Y	mass fraction

Greek

τ_{chem}	chemical time scale (ms)
η	combustion efficiency
ξ	cumulative spectral power
ε	dissipation rate (s⁻¹)
ρ	density (kg/m³)
μ	dynamic viscosity (Ns/m²)
ε_b	emissivity at the thermocouple bead surface
Φ	equivalence ratio
v	kinematic viscosity (m²/s)
$\propto_f$	mean fuel mass fraction at a particular plane
θ_0	momentum boundary layer thickness
ψ	percentage of maximum reverse flow velocity (%)
τ_{res}	residence time scale (ms)
τ^*	residence time of fine structure (in EDC model)
δ_ω	shear layer thickness (vorticity thickness)
ζ	spatial mixedness
ω	specific dissipation rate
σ_f	standard deviation in fuel mass fraction at a particular plane
σ	Stefan-Boltzmann constant $\left(5.67 \times 10^{-8} W / m^2 K^4\right)$
ΔP	total pressure drop (N/m²)

Subscripts

a	air
atm	atmosphere

c	cavity
C	combustor
3	station number for combustor inlet
4	station number for combustor exit
f	fuel
ms	mainstream
Ox	oxidizer
p	primary air
pr	product

REFERENCES

Atul, S. and Ravikrishna, R. V., Single Cavity Trapped Vortex Combustor Dynamics – Part-1: Experiments, 2011, *International Journal of Spray and Combustion Dynamics*, **Vol. 3**, pp. 23–24.

Bucher, J., Edmonds, R. G., Steele, R. C., Kendrick, D. W., Chenevert, B. C., and Malte, P. C., The Development of a Lean Premixed Trapped Vortex Combustor, *ASME Turbo Expo 2003*, Atlanta, 16–19 Jul 2003, (*GT-2003–38236*).

Danckwerts, P. V., The Effect of Incomplete Mixing on Homogeneous Reactions, 1958, *Chemical Engineering Science*, **Vol. 8**, pp. 93–102.

Driscoll, J. F. and Jacob, T., Role of Swirl in Flame Stabilization, *49th AIAA Aerospace Sciences Meeting*, Florida, 4–7 Jan 2011, (*AIAA 2011–108*).

Edmonds, R. G., Steele, R. C., Williams, J. T., Straub, D. L., Casleton, K. H., Bining, A. Ultra-Low NOx Advanced Vortex Combustor, *ASME Turbo Expo 2006, Spain*, 8–11 May 2006.

Edmonds, R. G., Williams, J. T., Steele, R. C., Straub, D. L., Casleton, K. H., and Bining, A., Low NOx Advanced Vortex Combustor, 2008, *Journal of Engineering for Gas Turbine and Power*, **Vol. 130**, pp. 034502-1–4.

Ezhil Kumar, P. K., Experimental and Numerical Studies of Twin Cavity 2D Trapped Vortex Combustor, 2014, *PhD Thesis*, Indian Institute of Technology, Kanpur, India.

Ezhil Kumar, P. K. and Mishra, D. P., Numerical Simulation of Cavity Flow Structure in an Axisymmetric Trapped Vortex Combustor, 2012, *Aerospace Science and Technology*, **Vol. 21**, pp. 16–23.

Ezhil Kumar, P. K., and Mishra, D. P., Combustion Noise Characteristics of an Experimental 2D Trapped Vortex Combustor, 2015, *Aerospace Science and Technology*, **Vol. 43**, pp. 388–394.

Ezhil Kumar, P. K., and Mishra, D. P., Flame Stability Characteristics of Two-Dimensional Trapped Vortex Combustor, 2016, *Combustion Science and Technology*, **Vol. 188**, pp. 1283–1302.

Ezhil Kumar, P. K., and Mishra, D. P., Combustion Characteristics of a Two-Dimensional Twin Cavity Trapped Vortex Combustor, 2017a, *Journal of Engineering for Gas Turbine and Power*, **Vol. 139**, pp. 071504–071514.

Ezhil Kumar, P. K. and Mishra, D. P., Numerical Study of Reacting Flow Characteristics of a 2D Twin Cavity Trapped Vortex Combustor, 2017b, *Combustion Theory and Modelling*, **Vol. 21**, pp. 658–676.

Ezhil Kumar, P. K. and Mishra, D. P., Characteristics of Turbulent Flow Past Passive Rectangular Cavity at Large Reynolds Number, 2018, *Journal of The Institution of Engineers (India): Series C*, **Vol. 99**, pp. 223–232.

Gutmark, E. J., Paschereit, C. O., Guyot, D., Lacarelle, A., Moeck, J. P., Schimek, S., Faustmann, T., and Bothien, M. R., Combustion Noise in a Flameless Trapped Vortex Reheat Burner, *13th AIAA/CEAS Aero-Acoustics Conference*, Italy, 21–23 May 2007, (*AIAA 2007–3697*).

Hendricks, R. C., Shouse, D. T., and Roquemore, W. M, Water Injection Turbomachinery, NASA-TM-2005–212632, 2005.

Hendricks, R. C., Shouse, D. T., Roquemore, W. M., Burrus, D. L., Duncan, B. S., Ryder, R. C., Brankovic, A., Liu, N.-S., Gallagher, J. R., and Hendrickss, J. A., Experimental and Computational Study of Trapped Vortex Combustor Sector Rig with High-Speed Diffuser Flow, 2001, *International Journal of Rotating Machinery*, **Vol. 7**, pp. 375–385.

Hsu, K.-Y., Goss, L. P., and Roquemore, W. M., Characteristics of a Trapped Vortex Combustor, 1998, *Journal of Propulsion and Power*, **Vol. 14**, pp. 1–12.

Hsu, K.-Y., Goss, L. P., Trump, D. D., and Roquemore, W. M., Performance of a Trapped Vortex Combustor, *33rd Aerospace Sciences Meeting and Exhibit*, Reno, 9–12 Jan 1995, (*AIAA 95–0810*).

Huellmantel LW, Ziemer RW, Cambel AB. Stabilization of Premixed Propane-Air Flames in Recessed Ducts Journal of Jet Propulsion. 27: 31-34, AIAA, Reston, USA.

Katta V. R. and Roquemore, W. M., Study on Trapped-Vortex Combustor-Effect of Injection on Flow Dynamics, 1998a, *Journal of Propulsion and Power*, **Vol. 14**, pp. 273–281.

Katta, V. R. and Roquemore, W. M., Numerical Studies on Trapped Vortex Concepts for Stable Combustion, 1998b, *Journal of Engineering for Gas Turbine and Power*, **Vol. 120**, pp. 60–68.

Kim, M. K., Yoon, J., Park, S., Lee, M.-C., and Yoon, Y., Effects of Unstable Flame Structure and Recirculation Zones in a Swirl-Stabilized Dump Combustor, 2013, *Applied Thermal Engineering*, **Vol. 58**, pp. 125–135.

Lefebvre, A. H. and Ballal, D. R., *Gas Turbine Combustion*, 3rd Ed., Taylor and Francis, 2010, Boca Raton.

Little Jr., B. H. and Whipkey, R. R., Locked Vortex after Bodies, 1978, *Journal of Aircraft*, **Vol. 16**, pp. 296–302.

Meyer, T. R., Brown, M. S., Fonov, S., Goss, L. P., Gord, J. R., Shouse, D. T., Belovich, V. M., Roquemore, W. M., Cooper, C. S., Kim, E. S., and Haynes, J. M., Optical Diagnostics and Numerical Characterization of a Trapped Vortex Combustor, *38th AIAA/ASME/SAE/ASEE Joint Propulsion Conference and Exhibit*, Indiana, 7–10 Jul 2002.

Mishra, D. P., *Fundamentals of Combustion*, Revised Ed., Prentice Hall of India, New Delhi, 2010.

Mishra, D. P., *Gas Turbine Propulsion*, MV Learning, NewDelhi, 2015.

Mishra, D. P. and Sudharshan, R., Numerical Analysis of Fuel-Air Mixing in a Two-Dimensional Trapped Vortex Combustor, 2009, *Proceedings of IMechE, Part G: Journal of Aerospace Engineering*, **Vol. 224**, pp. 65–75.

Muruganandam, T. M., Nair, S., Olsen, R., Neumeier, Y., Meyers, A., Jagoda, J., Lieuwen, T., Seitzman, J., and Zinn, B., Blowout Control in Turbine Combustors, *42nd Aerospace Sciences Meeting & Exhibit*, Reno, NV, 2004, (*AIAA 2004–0637*).

Muruganandam, T. M., Nair, S., Scarborough, D., Neumeier, Y., Jagoda, J., Lieuwen, T. C., Seitzman, J. M., and Zinn, B., Active Control of Lean Blowout for Turbine Engine Combustors, 2005, *Journal of Propulsion and Power*, **Vol. 21**, pp. 807–814,

Muruganandam, T. M. and Seitzman, J. M., Characterization of Extinction Events near Blowout in Swirl-Dump Combustors, *41st AIAA/ASME/SAE/ASEE Joint Propulsion Conference and Exhibit*, Arizona, 10–13 Jul 2005, (*AIAA 2005–4431*).

Roquemore, W. M., Shouse, D., Burrus, D., Johnson, A., Cooper, C., Duncar, B., Hsu, K.-Y., Katta, V. R., Sturgess, G. J., and Vihinei, I., Trapped Vortex Combustor Concept for Gas Turbine Engines, *39th AIAA Aerospace Sciences Meeting & Exhibit*, Reno, 8–11 Jan 2001, (*AIAA 2001–0483*).

Shivashankara, B. N. and Crouch, R. W., Noise Characteristics of a Can Type Combustor, 1977, *Journal of Aircraft*, **Vol. 14**, pp. 751–756.

Shouse, D. T., Trapped Vortex Combustion Technology, (PPT Presentation) MITE Workshop, 2000.

Stone, C. and Menon, S., Simulation of Fuel-Air Mixing and Combustion in a Trapped Vortex Combustor, *38th AIAA Aerospace Sciences Meeting and Exhibit*, Reno, 10–13 Jan 2000, (*AIAA 2000–0478*).

Straub, D. L., Casleton, K. H., Lewis, R. E., Sidwell, T. G., Maloney, D. J., and Richards, G. A., Assessment of Rich-Burn, Quick-Mix, Lean-Burn Trapped Vortex Combustor for Stationary Gas Turbines, 2005, *Journal of Engineering for Gas Turbine and Power*, **Vol. 127**, pp. 36–41.

Sturgess, G. J. and Hsu, K.-Y., Entrainment of Mainstream Flow in a Trapped-Vortex Combustor, *Proc 35th Aerospace Sciences Meeting & Exhibit*, Reno, 6–9 Jan 1997, (*AIAA Paper 97–0261*).

Tuncer, O., Kaynaroglu, B., Karakaya, M. C., Kahraman, S., Oksan, C. Y., and Baytas, C., Preliminary Investigation of a Swirl Stabilized Premixed Combustor, 2014, *Fuel*, **Vol. 115**, pp. 870–874.

Wu, H., Chen, Q., Shao, W., Zhang, Y., Wang, Y., and Xiao, Y., Combustion of Hydrogen in an Experimental Trapped Vortex Combustor, 2009, *Journal of Thermal Science*, **Vol. 18**, pp. 256–261.

Xing, F., Wang, P., Zhang, S., Zou, J., Zheng, Y., Zhang, R., and Fan, W., Experiment And Simulation Study on Lean Blow-Out of Trapped Vortex Combustor with Various Aspect Ratios, 2012, *Aerospace Science and Technology*, **Vol. 18**, pp. 48–55.

Xing, F., Zhang, S., Wang, P., and Fan, W., Experimental Investigation of a Single Trapped-Vortex Combustor with a Slight Temperature Rise, 2010, *Aerospace Science and Technology*, **Vol. 14**, pp. 520–525.

Yi, T. and Gutmark, E. J., Combustion Instabilities and Control of a Multi-Swirl Atmospheric Combustor, 2007, *Journal of Engineering for Gas Turbine and Power*, **Vol. 129**, pp. 31–37.

8 Supersonic Combustion Ramjet Technology

V. Ramanujachari

CONTENTS

8.1 INTRODUCTION

The Boeing 747, better known as the Jumbo Jet, was first flown in 1969. After half a century, the evolution of commercial aircraft has brought us increasingly comfortable, efficient and safe aircraft. However, the time spent on these journeys has not changed. The supersonic Concorde, which substantially cut transatlantic flight times, was affordable for a few. Nowadays, although there are plans underway to revive supersonic flight, many people are looking further ahead, towards the hypersonic plane: from Europe to Australia and back taking the time of just one working day. The supersonic flight of Mach 2 (the Concorde used to fly at approximately Mach 2) and the hypersonic flight of Mach 5 require different technologies.

Today, manned hypersonic flights are only possible with rocket propulsion, such as those used in space missions. In these cases, the spacecraft carries its own reserve of liquid oxygen for combustion, which gives them autonomy outside the atmosphere, but increases weight and volume. These rocket engines have been tested in projects such as the X-15 by USAF, which in October 1967 set the speed record for a manned, powered aircraft at Mach 6.7, or 7,274 km/h. The main alternative to rockets are ramjets, and more specifically their supersonic combustion variant: the scramjet. The advantage of this engine is that it works with atmospheric oxygen, which is compressed in the engine intake due to the aircraft's own flight speed. Projects such as the USA's NASP; the Japanese aerospace plane; India's Hyperplane; the Chinese aerospace plane; Zanger, Germany's STS–2000; France's Tu–2000, and Russia's projects ended as the scramjet propulsion power plants which need further improvement. However, these programs have led to the development of very efficient high speed ramjets/scramjets integrated with experimental flying vehicles as research demonstrators to fly around a Mach number greater than 5. Till date, flight tests have been conducted in the USA for demonstration of scramjets in integration with experimental flying vehicles X-43A [1], the hypersonic flight demonstrator program (HyFly) [2] and X-51A [3]. Similar projects of integrated hypersonic demonstrators are being developed in Russia, China, France (LEA) [4], India (HSTDV) [5] and many other countries. But, the flight tests are yet to be successful to make either a weapon system or a commercial aircraft. The X-51 technology could be applied to the development of hypersonic missiles, currently underway in the USA. Although military applications today seem to be the most prominent and viable in the short term, new concepts for passenger transport may emerge from these developments. A ten seater aircraft may likely be developed by the year 2040. If these kinds of speeds are needed to fly either a commercial aircraft or a weapon system, the power plant—viz., scramjet engine technology—has to mature to the highest level of reliable operation for a long duration.

8.2 HYPERSONIC FLOW

Hypersonic flow is occurring at higher Mach numbers where a number of flow effects such as high temperature gas effects, thin shock layers and viscous interaction come into effect. Though hypersonic flow does not correspond to an exact Mach number, the minimum value accepted is somewhere in the range of Mach number 5 to 7.

8.3 NECESSITY OF HIGH SPEED ATMOSPHERIC PROPULSION

Rockets can easily propel the objects at hypersonic speeds. As the propellants stored inside the rocket consume 80% of the total mass of the rocket, the payload mass fraction obtained will be around 4–5%. This makes the cost of launching the payload for the mission to be very expensive. As the interest here is to increase the payload mass fraction, air-breathing engines have become inevitable. Due to the utilization of atmospheric air for burning the fuel in the engine the mission cost is enormously reduced. In addition, from the energetics point of view, the air-breathing engines are far superior to rocket engines within the operating altitude as shown in Figure 8.1 [4]. It shows the specific impulse as a function of flight Mach number. Hydrogen fuel is considered for the ramjet and the scramjet systems. Though a turbojet delivers the highest specific impulse, its operating Mach number is very limited. The subsonic

Specific Impulse (sec)

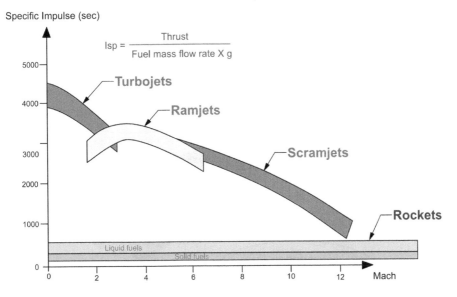

FIGURE 8.1   Specific impulse as a function of flight Mach number.

Source: Adapted with permission from Falempin, F. (2008).

combustion ramjets are inefficient for the flight Mach numbers greater than 3. The only option is the supersonic combustion ramjet for the flight Mach numbers from 3–10 with reasonable value of the specific impulse. The rockets deliver the lowest specific impulse, but are independent of the altitude of operation. This makes them attractive for interplanetary travel. Hence, the high speed air-breathing engine remains an important aspect of aerospace technology which can revolutionize space access and military capabilities.

8.4   HYPERSONIC VEHICLE/WEAPON SYSTEM

The superpowers—viz., USA, Russia and China—are actively engaged in the development of hypersonic weapons of at least Mach number 5 class with a purpose of striking prompt global targets and time critical targets. They could overcome detection and defense due to their high speed, maneuverability, survivability and low altitude of flight. They can be used as reusable launch vehicles for efficient space access. Figure 8.2 shows the details of a hypersonic air breathing vehicle.

This is an engine integrated with an airframe having wings and control surfaces. The bottom portion of the vehicle is the scramjet engine with its air intake, combustor and single expansion ramp nozzle. The idea of realizing this vehicle is to demonstrate the critical hypersonic technologies—viz., hydrocarbon-fueled supersonic combustor, endothermic fuel development for regenerative cooling of the engine, hypersonic aerodynamics and aero-thermodynamics, high temperature materials, hot structures and thermal protection system. The flight Mach number is in the range of 6 to 7. The corresponding total temperature will be 2000 K to 2500 K. The

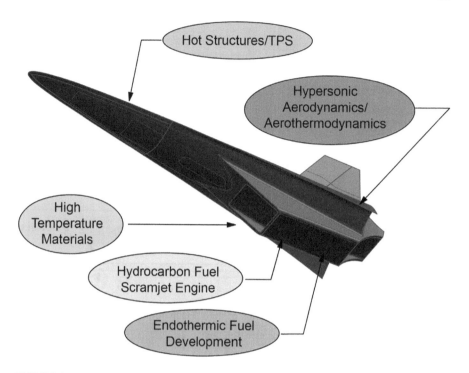

FIGURE 8.2 A typical hypersonic air-breathing vehicle.

altitude of cruise will be 30 to 35 km. Defense Advanced Research Projects Agency (DARPA), USA, with the support of the US Airforce (USAF) is continuing work on the hypersonic air-breathing weapon concept with the objective to develop critical technologies needed for air-launched hypersonic cruise missiles by 2023. Russia is developing Tsirkon, a ship-launched hypersonic cruise missile capable of flying at Mach number 8 with a range of about 1000 km, that can engage ground and naval targets, and the missile may be operational by 2023. China successfully tested Starry Sky–2 (Xing Kong–2), a nuclear capable hypersonic vehicle prototype in 2018. This vehicle has flown at Mach number 6 and is likely to be operational by 2025. India as a part of BrahMos-II intends to develop hypersonic cruise vehicle technologies to fly at Mach number 7 and the system may be operational by 2025–2028. France in collaboration with Russia is developing a hypersonic weapon under the project V-max Program. Other countries who have shown interest and are working on hypersonic programs are Iran, Israel and South Korea [6]. Hypersonic technologies are significant from the point of view of national security, transportation and space systems. The opportunities are to penetrate defenses, strike fleeting targets, agile targeting and high travel speeds. Challenges are heat tolerant materials, propulsion technology, limited testing resources and safety and control.

Figure 8.3 shows the variants of hypersonic air-breathing vehicle [7] providing applications of fast long-range civil transport around the globe, long-range cruise

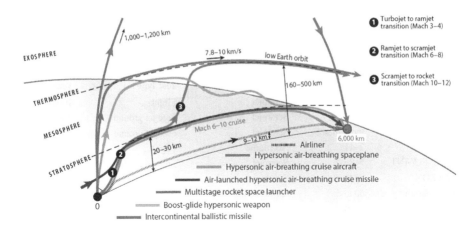

FIGURE 8.3 Variants of hypersonic air-breathing vehicle.

Source: Adapted with permission from Urzay, J. (2018).

missiles and high launch rate space transportation of payloads to LEO. As the air-breathing systems will be starved of air at higher altitudes, it is essential to have an additional rocket system to get out of the mesopause with a Mach number of the order of 20–25 to reach LEO altitudes with enough energy for injection of the payload. The air-breathing engine option provides lesser all up weight compared to a similar all rocket option. The challenge here is to capture the mass and add energy to a system flying at ultra-high speed—viz., 2–10 km/s. In addition, the engine has to generate a large amount of thrust to overcome the external and internal drag by the addition of energy within a fraction of a second residence time.

As shown in Figure 8.3 the airliner flies at an altitude of 9–12 km with a speed close to the sonic speed. The hypersonic space plane uses a combination cycle engine using turbojet, ramjet and scramjet for climb and cruise and payload injection by a solid propellant rocket. Cruise aircraft and cruise missiles can have many things in common except the initial launch platform. Cruise missiles will have an airdrop-solid booster and scramjet option. Whereas, the cruise aircraft can use a turbojet-ramjet-scramjet combination to start from the runway and reach the destination. These vehicles can fly around 30 km altitude, wherein atmospheric air is available for the engine operation. Though many nations are in the business of realizing hypersonic weapons of Mach 5+ speed, none has made an operational military weapon so far. A special mention needs to be made for a very successful hypersonic program carried out in the USA. X-51 is the AFRL's successful hypersonic vehicle program with a goal of providing sustained operation of a scramjet towards achieving long-duration flight. The X-51A scramjet engine used more practical endothermic hydrocarbon fuel which could cool the engine in a regenerative way and hence achieved extended duration. Similar to the X-43, it was an air-launched vehicle with a booster stage with a modified Army solid rocket booster (SRB) instead of a modified Pegasus launch vehicle. After numerous design iterations, ground tests and risk

TABLE 8.1
Summary of the Hypersonic Programs.

Name	Mach Number	Engine Type	Test Type	Year of End
Kholod	more than 6	Scramjet	Flight	1998
IGLA	6–14	Scramjet	Free jet ground	2009
X-2000	more than 5	Scramjet	Free jet ground	2011
SHEFEX	7–11	No engine	Flight	2011
HEXAFLY-INT	7–8	Scramjet	Free jet ground	Present
SKYLON	up to 16	Combined	Concept stage	Present
JAXA aircraft	up to 5	Air-breathing engine	Free jet ground	Present
14-X	6–10	Scramjet	Concept stage	Present
X-43A	up to 9.5	Scramjet	Flight	2010
X-51A	up to 5.1	Scramjet	Flight	2011
HIFiRE	up to 7	Scramjet	Flight	Present
SR-72	5	Combined	Concept stage	Present
CNUDT engine	4.5	Detonation ramjet	Free jet ground	Present
ISRO Scramjet	6.0	Scramjet	Captive flight	2016
HSTDV	6.0	Scramjet	Autonomous flight	2020

Source: Data used with permission from Arefyve, K.Y., et al. (2019).

reduction procedures, the X-51A flight vehicle conducted 210 seconds of scramjet powered flight, confirming the predicted long-range capabilities of scramjets. This success has paved the way for getting confidence in making a weapon system based on scramjet technology. The status of many scramjet programs are summarized in Table 8.1 [6].

8.5 SCRAMJET PROPULSION SYSTEM

Figure 8.4 shows the elements of a scramjet propulsion system. It consists of external and internal compression air intake, isolator and divergent combustor with fuel injection and flame stabilization devices and single expansion ramp nozzle. A brief description of the components are given next.

8.5.1 AIR INTAKE

Air intake compresses the air through multiple shock waves. Shocks depend on the intake geometry and flight Mach number. This in turn is responsible for the performance parameters such as mass flow rate, total pressure recovery (ratio between total pressure at the exit of the intake and the free stream total pressure) and temperature ratio (ratio between static temperatures at the end of the intake and the free stream). A hypersonic vehicle flying at a flight Mach number of 6.5 at an altitude of 35 km would produce a static pressure ratio of about 60 by decelerating the flow to a Mach number of 2.0 at the inlet of the combustor. The static temperature ratio would be

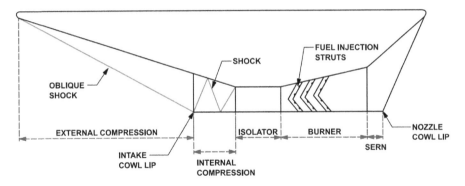

FIGURE 8.4   Elements of scramjet propulsion system.

of the order of 4.5. The stagnation temperature of the air entering the intake would be of the order of 2000 K. These values are required for the development of ground test facilities. The total pressure recovery reported for the hypersonic intake would be of the order of 15–20%. This is much less compared to a value of about 85% for a supersonic intake. Hypersonic air intakes are basically variable geometry intakes. At least one time change of geometry is needed for the starting of the intake as the contraction ratio may not be large enough to achieve self-starting of the intake.

8.5.2   Mixed Compression Air Intake

The intake configuration considered is a single ramp with cowl. The compression process is both in external and internal modes. The external compression is due to the forebody shock emanating from the nose of the intake, and the internal compression follows the external compression comprising of multiple shock reflections. The forebody shock graces the cowl lip. This results in capturing the maximum mass flow rate of air. This is the design condition and shown in Figure 8.5a.

Forebody shock interaction with cowl can be determined using gas dynamic relationships [8]. Based on the shock interaction, it is possible to determine whether the inlet is on-design or off-design condition. Figure 8.5b shows the subcritical operation of the intake. In this case, the partial capture of the air is taking place. There is a lot of mass spillage which contributes to the drag on the vehicle. In addition, the combustion chamber will be starved of air for its need to release the energy by mixing and ignition with the fuel. This off-design operation is detrimental to the overall operation of the engine. Figure 8.5c shows the operation of the intake at supercritical condition. Though there will not be any deficiency in ingesting the mass flow rate, the expansion wave–shock wave interaction affects the pressure recovery and accelerates the flow through expansion waves. Deceleration of the flow and static pressure build up are the basic requirements from the intake which will be compromised in this mode of operation with acceleration of flow downstream of expansion fans. Also, the total pressure loss increases as a result of interaction of the shock waves and expansion waves with the boundary layer. The next item as shown in Figure 8.5

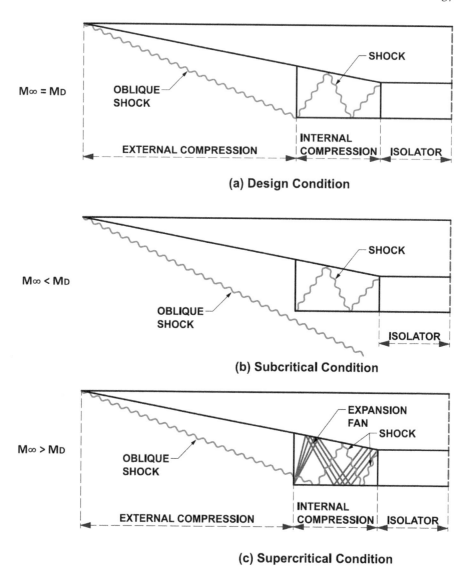

(a) Design Condition

(b) Subcritical Condition

(c) Supercritical Condition

FIGURE 8.5   A typical mixed compression air intake.

is to evaluate the internal compression at on-design and off-design conditions
($M_0 \leq M_{design}$). It involves multiple shock reflections inside a converging cross-section.
The flow after passing through the forebody shock is oriented along the ramp. The
first shock inside the internal compression duct originates from the cowl lip for
$M_0 \leq M_{design}$ which is a left running wave. The shock is constructed and the interaction with the ramp side is the reflection point. From this reflection point the next
wave would be the right running shock wave. Several left and right running shock
waves can be constructed until the final shock crosses the inlet throat section. This

shock modeling gives a two-dimensional flow field from nose to the throat. Beyond the throat complex interactions between expansion waves and shock waves take place. The formulation of the equations and solution details are available in standard textbooks on gas dynamics. A case study is given in Ref. [9].

8.5.3 ISOLATOR

As shown in Figure 8.6, it is a constant area duct [9] which provides adequate pressure rise to avoid the combustor pressure affecting the compression process in the air intake. Due to the supersonic combustion process downstream, the pressure increases in the combustor. If it is high, the flow separates in the isolator to accommodate the back-pressure rise through formation of pre-combustion shock waves. Based on the combustor conditions the isolator operates in three steady modes—viz., shock free mode, oblique shock train mode and normal shock wave mode.

The shock free isolator mode (Figure 8.6a) has a thin boundary layer growth along the walls producing a system of weak oblique shocks in the core flow. The pressure rise is of the same order as that of the raise predicted by the Fanno Flow [8]. The oblique shock train mode (Figure 8.6b) is seen when the combustor back pressure is large enough to separate the boundary layer inside the isolator. This results in blockage to the supersonic core flow by compressing it through a system of oblique shocks. When the subsonic combustion is encountered in the combustor, the normal

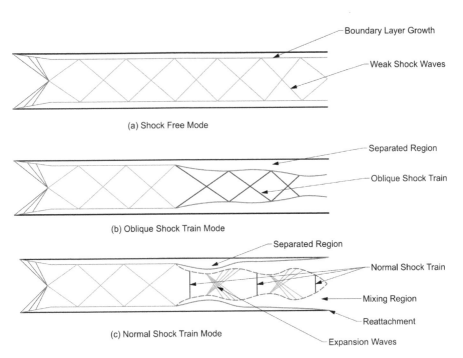

FIGURE 8.6 Schematic of isolator.

Source: Adapted with permission from Gopal, V. (2015).

shock train mode gets established in the isolator as shown in Figure 8.6c. Extensive studies are reported in Ref. [10].

8.5.4 COMBUSTOR

The combustor is a divergent combustor with an internal nozzle and is shown in Figure 8.7. There are 5 struts located in the divergent part of the combustor [11–13]. The idea here is to release energy at an appropriate divergent area to avoid thermal choking and combustor intake interactions. In addition, the fuel is spread fairly throughout the width of the combustor to achieve uniform mixing of the reactants. Otherwise, the high momentum air will flow along the combustor without mixing with the fuel leading to the presence of unburnt fuel and oxygen in the exhaust giving rise to poor combustion efficiency.

A typical static pressure profile is shown in Figure 8.8. The variations are plotted for reactive flow inside the combustor. Initially the wall static pressure increases due to heat release and after the combustion ceases the pressure decreases as a result of expansion through the divergent section [13]. As this static test is performed on the ground, in order to avoid separation of the flow in the combustor, a downstream ejector flow facility is necessary to impose the back-pressure condition as much less than the ambient pressure on ground. This is the back-pressure simulation corresponding to high altitude. This is one of the key components of the facility and its operation needs proper sequencing of the test to avoid malfunctioning of the ejector system.

In scramjet combustion two important characteristics are observed. In most of the configurations, the fuel and air are injected separately and then mixed. The velocities involved in fuel and oxidizer free streams are about two to three orders of magnitude greater than the premixed laminar flame velocity [7]. It does not mean that the scramjet combustion is totally non-premixed combustion. Premixed or partially premixed flames occur in the recirculation regions of the fuel injection cum flame

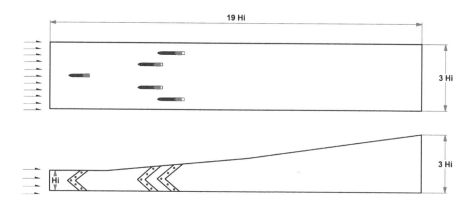

FIGURE 8.7   Schematic of a strut based combustor.

Source: Chandrasekhar, C. et al. (2012).

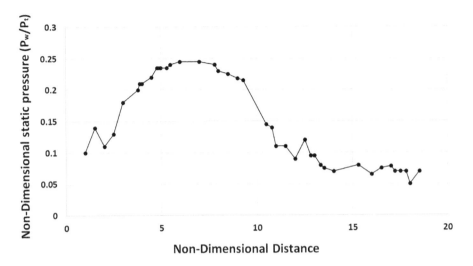

FIGURE 8.8 A typical static pressure profile along the combustor axial length.

Source: Chandrasekhar, C. et al. (2012).

holding devices. Mixing of fuel and air is important for the efficient heat release in the combustor as the compressibility effects inhibit these processes. Therefore, choice of injection system is of primary importance for the successful operation of the combustor. The fuel injector is expected to deliver the required amount of fuel to sustain combustion with minimum total pressure loss. The fuel selection is essential to the deliverable performance of a vehicle. Hydrogen was used in the programs— viz., X-43A, HIFiRE–3, HyShot-2, SCRAMSPACE, configurations of ONERA/ JAXA. JP-7 is used in X-51A; mixtures of ethylene with methane in HIFiRE-2. Silane is added for easy ignition in X-43A. Ethylene is used for igniting JP-7 and preheating it in X-51A.

When the isolator operates in shock free or oblique shock train mode, the combustor operates in supersonic combustion mode. Supersonic mixing involves semi-empirical relations based on experimental data which are adequate for engineering applications. The mixing efficiency is identified as a figure of merit for injection of fuel and oxidizer. A brief summary of the mixing models are presented next.

8.5.5 PARALLEL INJECTION MODEL

Example: Strut based injection system

$$\eta_m = \frac{x}{L_{mix}} \tag{8.1}$$

$$L_{mix} \cong 0.179 HC_m e^{1.72\varphi} \tag{8.2}$$

where x is the axial distance from the injector, φ is the equivalence ratio, H represents the scale of segregation. Two fluids such as fuel and oxidizer are mixed to a macroscopic scale, known as a segregated mixture, but not to the microscopic level. The entry scale of separation of these two streams is defined as the scale of segregation. The value of C_m, the mixing constant, typically varies from 25 to 60. An average value of $C_m = 46$ is considered [14].

8.5.6 PERPENDICULAR HYDROGEN INJECTION MODEL

Example: Normal injector

$$\eta_m = 1.01 + 0.176 \, ln\left(\frac{x}{x_\phi}\right) \tag{8.3}$$

$$x_\phi = 0.179 L_{mix} exp\left(0.172\phi\right) \tag{8.4}$$

$$L_{mix} \approx 60G \tag{8.5}$$

where G represents the spacing of normal injectors. Note that this mixing model is only valid for hydrogen based fuel injection [15].

8.5.7 STRUT MIXING MODEL FOR HYDROGEN

$$\eta_m = a\left(1 - e^{-\left(\frac{kx}{L_{mix}}\right)^d}\right) \tag{8.6}$$

$$L_{mix} = \frac{D_f K^*}{f\left(M_c\right)}\left[\frac{\rho_f u_f}{\rho_a u_a}\right]^{1/2} \tag{8.7}$$

The constants are given by $a = 1.06492$, $k = 3.69639$ and $d = 0.80586$. The term $f(M_c)$ is a compressible correction factor given by $f\left(M_c\right) = 0.25 + 0.75e^{-3M_c^2}$. The fuel jet diameter or thickness is given by D_f. The convective Mach number M_c is given by:

$$M_c = \frac{u_f - u_a}{a_f + a_a} \tag{8.8}$$

where u_f and u_a are the fuel injection velocity and the air velocity respectively. The acoustic velocity of air and fuel are given by a_a and a_f respectively. For slot type injection, $K^* = 390$ and varies with different injection configurations. Note that this mixing model is only valid for hydrogen fuel injection [16].

8.6 ANALYSIS OF FLOW THROUGH THE COMBUSTOR

The flow field in the burner can be computed by the quasi one-dimensional equations of mass continuity, axial momentum and energy equations. The driving potentials for the variation of properties such as pressure, temperature, density, velocity, Mach number and stream function are area variation, friction, drag due to immersed bodies like injectors and flame holding devices, variation in total temperature due to combustion and variation in mass flow rate due to additional injection of fuel at different locations in the burner. Due to the combined effect of these driving potentials, the variation in flow Mach number is given by the following equation:

$$\frac{dM}{M} = \frac{\left(1 + \frac{\gamma - 1}{2} M^2\right)}{1 - M^2} \left\{ \frac{dA}{A} + \frac{\gamma M^2}{2} \left[\left(\frac{4\mathfrak{f} dx}{D}\right) + \frac{2\delta D}{\gamma M^2 pA} \right] \right.$$
$$\left. + \frac{\left(1 + \gamma M^2\right) dT_t}{2} + \left[\left(1 + \gamma M^2\right) - y\gamma M^2 \right] \frac{d\dot{m}}{\dot{m}} \right\} \tag{8.9}$$

where M: Mach Number; A: Area; $4\mathfrak{f}$: Darcy's friction coefficient; x: axial distance; p: static pressure; T_t: total temperature; y: velocity ratio; y = 1 for parallel injection; y = 0 for perpendicular injection; $\dot{m}$: mass flow rate; γ: ratio of specific heats; δD: internal drag force caused by the immersed bodies; D: hydraulic mean diameter; d refers to the change.

This equation can be solved using Runge Kutta Order IV method for a given inlet condition and known geometry. Once Mach number distribution is computed, other flow variables can be calculated by standard gas dynamic equations. The details are given in Ref. [8]. The mixing efficiency, η_m, can be used in the mass continuity equation [9] as follows:

$$\frac{1}{\dot{m}} \times \frac{d\dot{m}}{dx} = \frac{f}{\left(1 + f\eta_m\right)} \times \frac{d\eta_m}{dx} \tag{8.10}$$

where f is the fuel air ratio.

The Mach number reduces along the length due to the variation in stagnation temperature, friction, mass addition and drag forces (occlusion of the flow). Only the area variation accelerates the flow leading to increase of Mach number (relieving of the flow). As long as the heat addition is predominant, the Mach number decreases and tends towards thermal choking. It does not attain unity Mach number as the quantity dM/M tends to infinity. One can design the combustor in such a way that the effect of area variation becomes predominant to accelerate the Mach number before attaining thermal choking. This implies that in the one-dimensional sense the entry supersonic Mach number attains transonic value and increases to supersonic value in the internal nozzle. The computation of total temperature at every incremental axial distance used for numerical integration can be carried out based on chemical equilibrium using the NASA CEC-71 software package [17]. A combustion efficiency will have to be considered during the computation of temperature distribution, as the calculation of adiabatic flame temperature assumes 100% combustion efficiency, which is not practical in the scramjet combustor. The combustion efficiency is always less than or equal to mixing efficiency as the micro mixing has to take place before combustion commences. One of the definitions of combustion efficiency reported in Ref. [14] is given next:

$$\eta_b = \left(T(x) - T_{in}\right)/\left(T_{aft,in} - T_{in}\right) \qquad (8.11)$$

where T(x) is the axial distribution of static temperature along the combustor. T_{in} is the mass averaged static temperature of both fuel and air streams at burner entry. $T_{aft,in}$ is the adiabatic flame temperature where the reactant composition and static enthalpy are the mass averaged values of fuel and air streams at combustor entry. The value of dT_t/dx is greatest shortly after ignition, usually near the burner entry and decreases monotonically to its least value at the burner exit. Hence, the total temperature ratio can be written as follows:

$$\frac{T_t(x)}{T_{t,in}} = 1 + (\tau_b - 1)\frac{\theta\chi}{\left(1 + (\theta-1)\chi\right)} \qquad (8.12)$$

where τ_b is the overall total temperature rise ratio in the burner. $\chi = (x - x_i)/(x_b - x_i)$; where x_i is the axial location at which the supersonic combustion begins; x_b: combustor exit. θ is the empirical constant of the order of 1 to 10. This value depends on mode of injection and fuel air mixing [14]. This information can be utilized for the simple analysis of the combustor.

8.7   SINGLE EXPANSION RAMP NOZZLE

The nozzle expands the product gases from the combustor exit and accelerates the flow to produce necessary forward thrust. It consists of a ramp and the cowl as shown in Figure 8.9a. A series of Prandtl-Mayer expansion fans originate from the combustor exit which are then reflected from the nozzle cowl and finally get cancelled at the ramp surface of the nozzle. The expansion process is complete at the cowl lip when the nozzle is operating at on-design and the slip line between internal flow and external flow is along the undisturbed external flow. The over expanded nozzle is shown in Figure 8.9b. The over expanded nozzle mode is established when the ambient pressure is greater than the design point nozzle exit pressure calculated by the expansion process at design point. The slip line between internal and external flow is established by equating nozzle exit pressure and the ambient pressure at the nozzle exit. As the slip line is in the upward direction, expansion fans originate from the corner for the hypersonic oncoming external flow. The under expanded nozzle is shown in Figure 8.9c. This mode is established when the ambient pressure is less than the design point nozzle exit pressure. The slip line between internal and external flow is found out by matching the nozzle exit pressure and external pressure. As the slip line is downward the oblique shock is formed by the hypersonic flow turning the corner. Case studies are reported in Ref. [9].

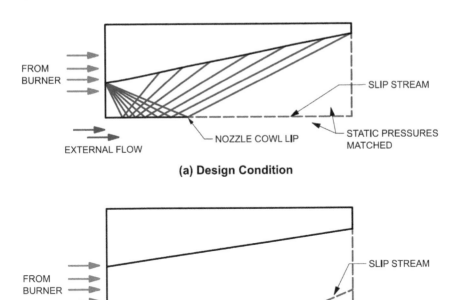

(a) Design Condition

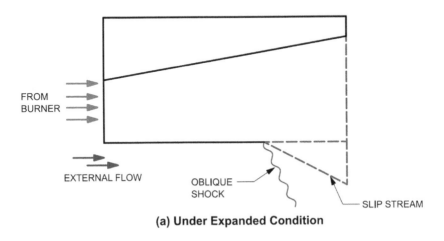

(a) Over Expanded Condition

(a) Under Expanded Condition

FIGURE 8.9   Schematic of a single expansion ramp nozzle.

8.8   TECHNOLOGICAL CHALLENGES

Some of the technological challenges are briefly explained as follows:

1. It is a challenge to integrate different propulsion systems to boost the scramjet above its starting Mach number as it cannot develop static thrust. This needs additional structural members which can increase the weight leading to a lower payload mass fraction. It is essential to understand the aerothermal environment with which the hypersonic vehicle is flying. Heat loads are very high at hypersonic speeds as it is proportional to the third power of the flight velocity. In the vehicle different components are subjected to different heat fluxes. Hence, suitable materials will have to be selected for different sections to withstand the thermomechanical loads. Shock boundary layer interactions produce flow separation and reattachment. These regions are vulnerable to heating. It is essential to have better predictive capability using CFD or an experimental database to overcome the thermal failures in such regions.

2. In the intake/isolator, starting, air mass capture, stagnation pressure recovery, contraction ratio are the important aspects to be considered. Starting is a process of establishing supersonic flow in the internal compression part of the intake. The contraction ratio is predominantly responsible for starting of the intake. To determine a boundary of an allowable contraction ratio, the Kantrowitz limit is widely used. This is determined for a thermally and calorically perfect gas with the assumption that a normal shock, which appears in the supersonic diffuser, would be pushed upstream towards the throat under back-pressure conditions and would allow the inlet to remain started as long as the normal shock stays within the diffuser; in the limit, the normal shock will be at the throat. Based on quasi-one-dimensional flow, the inverse of contraction ratio in the Kantrowitz limit [15] is

$$\frac{A_2}{A_{throat}} = \frac{1}{M_2} \left[\frac{(\gamma+1)M_2^2}{(\gamma-1)M_2^2+2} \right]^{\frac{\gamma}{\gamma-1}} \left[\frac{\gamma+1}{2\gamma M_2^2-(\gamma-1)} \right]^{\frac{1}{\gamma-1}}$$

$$\left[\frac{1+\left(\dfrac{\gamma+1}{2}\right)M_2^2}{\dfrac{\gamma+1}{2}} \right]^{\frac{\gamma+1}{2(\gamma-1)}} \tag{8.13}$$

where station "2" stands for inlet to the internal compression intake. Here, A: Area; M: Mach number; γ: ratio of specific heats. If the flow is assumed to remain isentropic throughout the compression process, the area ratio derived from continuity equation provides another limit. The inverse of contraction ratio is

$$\left(\frac{A_o}{A_{throat}}\right)_{isentropic} = \frac{1}{M_o}\left[\frac{2}{\gamma+1}\left(1+\frac{\gamma-1}{2}M_o^2\right)\right]^{\frac{\gamma+1}{2(\gamma-1)}} \qquad (8.14)$$

where "o" stands for free stream condition. These functions indicate that the contraction ratio increases with an increasing Mach number. Practical inlets operate between these two limits, and the ability to pass the entry mass flow through the inlet throat depends on a number of factors—viz., boundary-layer thickness, momentum distortion at the throat cross-section and the use of an inlet bleed.

As the contraction ratio for a realistic vehicle is very much less than the corresponding contraction ratios for self-starting, there is a requirement to change the area ratio in a variable way using a mechanism operated by an actuator at the beginning of the scramjet operation before ignition. This implies that the operating point lies between the Kantrowitz limit and the isentropic flow value as shown in Figure 8.10 [18].

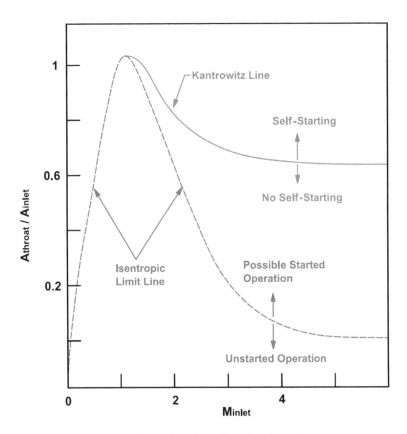

FIGURE 8.10   Contraction ratio as a function of inlet Mach number.

Source: Adapted with permission from Paniagua, G. (2014).

3. The schematic of the rotating cowl of the PREPHA engine air intake is shown in Figure 8.11 [4]. This is also a critical component as it can interfere with the oncoming flow which experiences shock boundary layer interactions. This can prevent the starting and leading to spillover of the flow. The spillover is detrimental to the flow as it increases the drag on the vehicle needing more thrust to be produced by the propulsion system. As the thrust is directly proportional to the mass captured by the intake, there will be deficiency in the thrust leading to lower values of thrust margin (thrust-drag) or no margin for the forward motion of the vehicle.

   However, this rotating cowl is needed for starting of the air intake. At the time of injection of the hypersonic vehicle from the launch vehicle the air intake will be under a closed condition. Then the air intake is opened using an actuator mechanism. Essentially, the contraction ratio decreases as the intake area is increased during opening keeping the throat area constant. If this design contraction ratio is in between the Kantrowitz limit and the isentropic limit, then the air intake starts smoothly. The position of the intake cowl hinge point, time of opening of the rotating cowl and the locking of the cowl without backlash are critical to the reliable operation of the opening of intake cowl.

4. In the internal and external compression air intake, there is substantial stagnation pressure loss due to shock phenomenon. In addition, to overcome the starting process of air intake and cater for the large variation in flight speed of the vehicle, there is a need to operate it as variable geometry intake [4]. This increases the complexity of design and hardware implementation and poses a challenge to the successful and repeatable operation of the intake.

5. The combustion process is a technological barrier of the highest order, wherein the supersonic fuel air mixing is impaired by compressibility

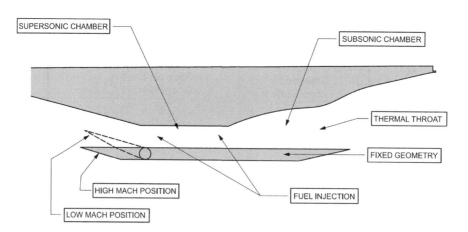

FIGURE 8.11   Schematic of the rotating cowl of PREPHA engine air intake.

Source: Adapted with permission from Falempin, F. (2008).

effects which does not permit efficient combustion to occur in a short time of the order of 2 milliseconds. Unless the micro mixing of the flow takes place, there is no combustion possible. The other factors are complex transport phenomena, chemical kinetics, turbulence, flow recirculation, shock induced combustion, thermal choking and near wall burning.

6. Ignition delay is a road block to efficient combustion as the flow and chemical time scales are of the same order. The auto-ignition of Jet-A1 type of hydrocarbon fuel combustion needs the temperature of the order of 1000–1400 K. This temperature has to come from decelerating the free stream hypersonic flow to supersonic flow inlet to the combustor through shocks. The ignition time is of the order of 10–100 microseconds. In most of the situations using hydrocarbon fuels, the ignition delay is two orders of magnitude greater than flow residence time. Use of fuels like JP-7 and JP-10 are being tried for better ignition and combustion. At the same time their endothermic characteristics are utilized for engine cooling. Yet another challenge is the gasification, storage and injection of fuel.

7. Fuel injection and flame holding are the challenges to complete the combustion inside the combustion chamber. Fuel injection struts [12], cavities [19], ramps [20], pylons [21] are used for this purpose. As they are introduced inside the flow they also provide flame stabilization due to recirculation regions created by their obstruction to the supersonic flow. These regions provide a continuous heat source for the oncoming reactive mixture to keep the reaction zone within the combustor. In order to depict this point, a typical strut injection system is shown in Figure 8.12 [11].

   As this strut is immersed inside the reactive flow region, it needs cooling. The liquid hydrocarbon fuel is used to circulate inside the cooling passage of the strut in two passes. In the first pass of the coolant the leading edge of the strut is cooled and in the second pass the fuel is injected through very tiny holes. As the fuel goes through two passes inside the strut it is heated before it enters the combustor, facilitating better ignition and combustion.

8. As the heat flux increases as the third power of flight velocity, the design and realization of the thermal protection system (material choice and structural design) becomes a challenge. Accurate prediction of temperature considering the wall boundary layer interaction remains a scientific barrier for realizing an operating hypersonic system. For short-duration experiments one can use the materials of construction without cooling for the scramjet system. If the duration goes more than 20 seconds, active cooling will be required. The liquid hydrocarbon fuels can be used as a coolant for the combustor and the fuel injection elements—viz., struts, pylons and short ramps. But, the coolant/fuel may not be adequate for the long-duration operation.

9. Yet another problem is to cool the air frame of the vehicle for a long duration. In this case use of heat-resistant materials with sufficient mechanical properties must be resorted to as the design would be based on hot structures. Presently the material scientists are coming out with exotic

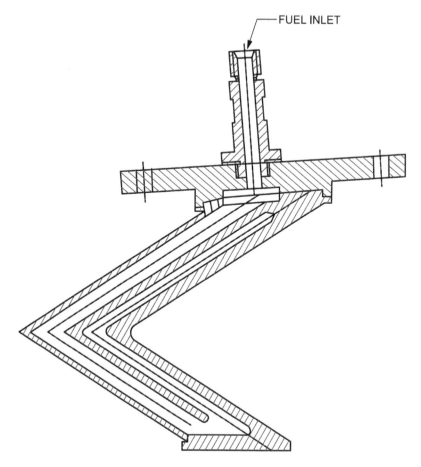

FIGURE 8.12 Schematic of the strut injector.

Source: Chandrasekhar, C., et al. (2014).

functionally graded materials, metallic foams, carbon composites and
refractory metals, niobium and nimonic alloys. The challenges of scramjet
combustion would be shifted to challenges in material science, if the hypersonic vehicle is to succeed. Endothermic fuels like JP-7 are being tried
out to cool the engine when the fuel absorbs the heat due to endothermic
decomposition and produces smaller chain hydrocarbons from long chain
multi-component fuels. This is an active area of research to get correct
fuels for long-duration flight.

This implies that the scramjet propulsion is the multidisciplinary area involving
interactions of structures, fluid dynamics, heat transfer, heat release, control, etc.
All these areas are yet to be understood and exploited for the realization of an operating weapon system.

8.9   SCRAMJET ENGINE REALIZATION

The flow path design of the engine can be carried out using a simple approach already explained. More detailed computations can be carried out using CFD. The high fidelity computations provide the ideas on injector locations and the amount of fuel to be injected from each orifice to get maximum combustion efficiency. It can also provide the values of frictional forces experienced by the injector elements and walls of the combustor, fuel rich and lean regions, temperature profiles, average Mach number profiles and shock structure. Some of these values can be used to carry out thermos-structural design of the various components of the engine, in addition to the computation of propulsion performance parameters such as thrust and specific impulse. Special heat-resistant materials such as nimonic or niobium can be considered for the combustor as it is subjected to aerodynamic heating on the wall outside and combustion heating on the wall inside. Here, it is essential to consider the buckling of the engine structure and addition of stiffeners to overcome it. The internal compression intake and the internal divergent nozzle can be part of this engine. This total engine component will be fitted to the airframe. The engine-airframe integration is a critical process as the forces have to be transferred in between the engine and the airframe. Close tolerances are needed in fabrication of the components as the flow encountered is supersonic inside the flow path and hypersonic outside the airframe. Provision will have to be made for the expansion of the components during heating in the course of a static test or the flight. The joining method of components needs attention to reduce dimensional distortions. Unnecessary protrusions will become points of emanating shocks and hotspots. Manufacturing of fuel injection elements with cooling like the strut shown in Figure 8.12 is very critical. Material selection, methodology of fabrication, avoidance of leakage are some of the issues to be addressed as it has to get immersed in the reactive flow and still perform the function of constant fuel injection for a particular duration and save the component from burnout.

8.10   TESTING OF THE SCRAMJET ENGINE

The testing of the engine has to be done in two stages—viz., ground testing and flight testing. All the known unknowns will be understood in ground tests. The design either carried out by CFD or empirical/one-dimensional models needs to be validated in the ground test facilities to assess the performance of the engine and the methodology used to design the system with certain assumptions. As the scramjet engine consists of an air intake, isolator, injection system, combustion chamber and internal nozzle, each component test has to be conducted in an isolated way to evaluate their individual performance. Then the integrated tests are to be conducted. Depending upon the availability of test facilities, the tests are to be carried out. To carry out reliable ground tests at hypersonic speeds, there are four key parameters required for the test facilities. They are, (i) testing gas to ensure correct thermo-chemistry in the flow; (ii) total pressure and total temperature of flow to be duplicated in such a way that the velocity of the flow has to be reproduced with proper facility nozzle flow expansion. Consequently, the chemical reaction process can be simulated correctly, as well as the aerodynamic forces and moments; (iii) the facility nozzle size at the exit must be big enough to accommodate a large vehicle model so that the chemical reaction time scale can be much smaller than that of the test gas flowing past the model; (iv) test duration should be long enough to sustain stable supersonic combustion in the engine. The challenge here is to simulate all four parameters in a single test run.

Methods of ground testing recommended towards realization of a scramjet propulsion system are: connected pipe, semi free jet and free jet [22]. In the connected pipe tests the isolated combustor or combustor with nozzle would be studied for evaluating the propulsion performance and thermo-structural integrity of the component. The combustor inlet conditions are simulated using a heater to produce high enthalpy air. This is expanded through a convergent-divergent facility nozzle to get the required Mach number at entry to the combustor. The heater is operated at a higher pressure to get the required static pressure inlet to the combustor. The typical values of entry to the combustor for a flight Mach number of 6 to 7 would be static temperature, 1100–1500 K; static pressure, 0.5–0.8 bar (abs); flow velocity, 1200–1600 m/s; Mach number, 2–2.5; and total temperature, 2000–2500 K. A hydrogen based heater can be used with air and excess oxygen to ensure 23% oxygen by mass in the products of combustion discharged through the facility nozzle to the combustor. The required amount of fuel is injected and mixed with high enthalpy vitiated air for getting auto-ignition and subsequent combustion. If auto-ignition is not possible, a separate ignition system can be used. An air ejector facility can be installed and operated downstream of the combustor to simulate the back pressure corresponding to the altitude of operation. This avoids flow separation in the combustor. The connected pipe facility can be used for the simulation of higher flight Mach numbers bypassing the air-intake losses and flow complexities. The shortcomings of this test would be (i) flow distortion effects cannot be simulated, (ii) inlet-combustor interactions cannot be studied and (iii) flow vitiation makes the performance predictions inaccurate, especially the presence of water vapor and carbon dioxide in the test gas leading to different chemistry in the ground test compared to the flight. Integrated flow path or engine testing can be done by semi free jet test methodology. The engine module is fully replicated, but the external intake and nozzle parts are either partially or not at all replicated. This approach is useful for testing engines of larger size. This methodology has been followed extensively in the USA [22]. The free jet testing involves testing the whole vehicle from tip to end. As the resources needed are enormous, either scaled tests or small vehicle tests can be done with this approach. For example, the hypersonic research engine, USA; scramjet engine, CIAM, Russia [22]; and X-43 full vehicle, USA were tested using this method.

Flight test is the ultimate one in the process of developing and proving an integrated system. It is intended to validate the design, mathematical analysis, CFD analysis and predictive tools exclusively developed for various sub-systems. In addition, the flight test reveals the structural integrity of the vehicle and its associated components under different altitudes and flight speeds. Normally for short-duration operation the hot structure design is carried out. Cooling of the air frame and engine structures is necessary for long-duration operation. This will be validated in the intended flight envelope and provides the designer the confidence on the margins available for extending the flight operations exceeding the bounds of the envelope. In air-breathing hypersonic vehicle flight, there are two variants of flight tests—viz., captive-carry and autonomous flights. Out of the two, the captive-carry is relatively simple as the hypersonic cruise vehicle (scramjet engine) is always attached to the launch vehicle. A sounding rocket can be used as a launch vehicle and the scramjet engine can be fired once the intended Mach number and altitudes are reached. The data can be acquired by a telemetry station on the ground. ISRO, India, has done this type of flight test to evaluate the hydrogen-fueled scramjet engine. In the autonomous mode, initially the hypersonic cruise vehicle is attached to the launch vehicle. The launch vehicle can be a single stage solid/liquid rocket or a combination of air drop and a small solid/liquid booster rocket launched to the required Mach number and altitude conditions. After the launch phase is over, the hypersonic vehicle will be ejected out and the scramjet engine is lighted to produce the thrust for the cruise phase. The autonomous mode flights have been carried out in the USA for X-43A and X-51A programs and for the HSTDV program of India. The telemetry data can be acquired by a ground station or a separate aircraft having extensive instrumentation analyzed.

8.11 CONCLUSION

The scramjet engine is uniquely suited for hypersonic propulsion. Though there are no moving parts like turbo machines, the system becomes very complex as the processes are controlled by aerodynamics and thermodynamics. External vehicle drag and heat loads are varying as a function of square and cube of flight speed respectively. This implies that a lot of drag has to be overcome and a lot of heat flux has to be handled during the flight. Therefore, getting a thrust margin and choosing materials for airframes and engines are the real challenges in this system. The design methodologies are well proven with the advent of maturity in CFD and availability of high-power computers. However, the ground tests and flight tests are very complex due to huge investments and limitations of the facilities in their inability to simulate the actual flight corridor. In addition, as scaling laws are not known, there is a need to do tests on the full-scale engine. A cost-benefit analysis is needed to embark on the scramjet technology development program as the investments on test facilities, test article realization, test facility and test article instrumentation, quality control and safety are huge. But, in a single program a lot of challenges are to be overcome in terms of scramjet development and the launch vehicle development and their integration and successful separation during flight and sustained scramjet flight acquiring thrust margin. This is truly a frontier area of aeronautics and there is an urgency to conquer it.

REFERENCES

1. Voland, R. T., Huebner, L. D. and McClinton, C. R., "X-43A Hypersonic Vehicle Development", *Acta Astronautica*, Vol. 59, pp. 181–191, 2006.
2. HyFly—Hypersonic Flight Demonstration, www.onr.navy.mil.

3. Rondeau, C. M. and Jorris, T. R., "X 51 Scramjet Demonstrator Program: Waverider Ground and Flight Test", STEP Southwest Flight Test Symposium, 28 Oct.–1 Nov. 2013, Ft Worth, T8.

4. Falempin, F., "Ramjets and Dual Mode Operation", In Advances on Propulsion Technology for High Speed Aircraft, pp. 7.1–7.35, Educational Notes, RTO-EN-AVT-150, 2008.

5. Rajinikanth, B. and Anavaradam, T. K. G., "Tip to End Simulation of Airframe Integrated Scramjet Engine", Eighteenth International Symposium on Airbreathing Engines, p. 1132, 2007.

6. Arefyve, K. Y., Kushinov, N. V. and Prokrov, A. N., "Analysis of Development Trends of Power Units for High Speed Flying Units", *Journal of Physics: Conference Series*, p. 1147, 2019.

7. Urzay, J., "Supersonic Combustion in Air Breathing Propulsion Systems for Hypersonic Flight", *Annual Review of Fluid Mechanics*, Vol. 50, pp. 593–627, 2018.

8. Zucrow, M. J. and Hoffmann, J. D., "Gas Dynamics", Vol. 1, John Wiley, New York, 1976.

9. Gopal, V., "Reduced Order Analysis of Dual Mode Supersonic Combustion Ramjet Propulsion System", M.S. Thesis, University of Texas at Arlington, December, 2015.

10. Emami, S., Trexler, E. A., Auslender, A. H. and Weidner, J. P., "Experimental Investigation of Inlet-Combustor Isolators for Dual Mode Scramjet at a Mach Number of 4", NASA Technical Paper-3502, 1995.

11. Chandrasekhar, C., Ramanujachari, V. and Kishan Kumar Reddy, T., "Evaluation of Kerosene Fueled Scramjet Combustor using a Combination of Cooled and Uncooled Struts", *Defence Science Journal*, Vol. 64, No. 1, pp. 5–12, 2014.

12. Chandrasekhar, C., Ramanujachari, V. and Kishan Kumar Reddy, T., "Experimental Investigation on the Effect of Total Temperature on the Performance of Strut Based Combustor Using Kerosene Fuel", National Propulsion Conference-2013, Paper No. 24016.

13. Chandrasekhar, C., Ramanujachari, V. and Kishan Kumar Reddy, T., "Experimental Investigations of Hydrocarbon Fueled Scramjet Combustor by Employing High Temperature Materials for the Construction of Fuel Injection Struts", *International Journal of Science and Technology*, Vol. 1, No. 12, pp. 671–678, 2012.

14. Heiser, W. H. and Pratt, D. T., "Hypersonic Airbreathing Propulsion", AIAA Education Series, AIAA, Washington, DC, 1994.

15. Segal, C., "The Scramjet Engine: Processes and Characteristics", Cambridge Aerospace Series, Cambridge University Press, Cambridge, 2009.

16. Driscoll, J. F., Huh, H., Yoon, Y. and Donbar, J., "Measured Lengths of Supersonic Hydrogen-Air Jet Flames Compared to Subsonic Flame Lengths and Analysis", *Combustion and Flame*, Vol. 107, No. 1, pp. 176–186, 1996.

17. Gordon, S. and Mcbride, B., "Computer Programme for Calculation of Complex Chemical Equilibrium Compositions", NASA-CEC, 1971.

18. Paniagua, G., Iorio, M. C., Vinha, N. and Sousa, J., "Design and Analysis of Pioneering High Supersonic Axial Turbines", *International Journal of Mechanical Sciences*, Vol. 89, pp. 65–77, 2014.

19. Mica, D. J. and Driscoll, J. F., "Combustion Characteristics of a Dual Mode Scramjet Combustor with Cavity Flame Holder", *Proceedings of the Combustion Institute*, Vol. 32, pp. 2397–2404, 2009.

20. Curren, E. T. and Murty, S. N. B., "Scramjet Propulsion", *Progress in Aeronautics and Astronautics*, Vol. 189, 2000.

21. Freeborn, A. B., Gruber, M. R. and King, P., "Swept Leading Edge Pylon Effects on a Scramjet Pylon-Cavity Flame Holder Flow Field", *Journal of Propulsion and Power*, Vol. 25, No. 3, pp. 571–582, 2009.

22. McClinton, C. R., "High Speed/Hypersonic Aircraft Propulsion Technology Development", In Advances on Propulsion Technology for High Speed Aircraft, pp. 1.1–1.32, Educational Notes, RTO-EN-AVT-150, 2008.

9 Advances in Gel Propellant Combustion Technology

Manisha B. Padwal and Debi Prasad Mishra

CONTENTS

9.1 INTRODUCTION

Conventional rocket propulsion systems employ liquid and solid propellants to generate thrust. A hybrid combination of propellants using solid fuels and liquid oxidizers is supposed to combine the advantages of the respective propellants into one propulsion system [1]. Performance of this relatively less common propulsion concept is still constrained at present [1]. At the same time, well-established liquid and solid propulsion systems are also limited in some areas by the constraints imposed by their respective physical states. On this background, still another combination of the properties of liquid and solid propellants into single propellant is possible once the conventional liquid propellants are gelled. In gelation, suitable gelling agents are mixed with liquid propellants to transform propellants into solid state. The present chapter is devoted to a discussion of the advances in gel propellant combustion. Topics are arranged to enable the discussion of essential elements of gel propellant combustion involving the propellant development, characterization, atomization, and combustion. Each of these four elements is described using pertinent research reports with emphasis on the new results obtained on novel hardware

DOI: 10.1201/9781003049005-9

and their importance to other elements. An in-depth survey of the field and discussion of many scientific and technological aspects of gel propulsion can be found in literature [2].

9.2 PROPELLANT DEVELOPMENT

The final gel propellant formulation is obtained after consideration of many competing and sometimes conflicting requirements of mission goals and achievable performance. We would illustrate the propellant development phase using the examples of cryogenic and hypergolic liquid propellants, two of the most common liquid propellants in rocket engines.

9.2.1 GELATION OF CRYOGENIC PROPELLANTS

Cryogenic liquid propellants are the mainstay of liquid rocket engines [3]. Highly complex engineering and elaborate handling guidelines must be employed for harnessing these propellants. Efforts have been made since the beginning to reduce at least part of the complexity by means of novel gelation techniques [4–8].

The case of cryogenic gel propellants is unique in that their development begins with the development of gellant itself. Many factors are responsible for this approach. As depicted in Figure 9.1, cryogenic propellants (almost exclusively LH_2 and LO_2)

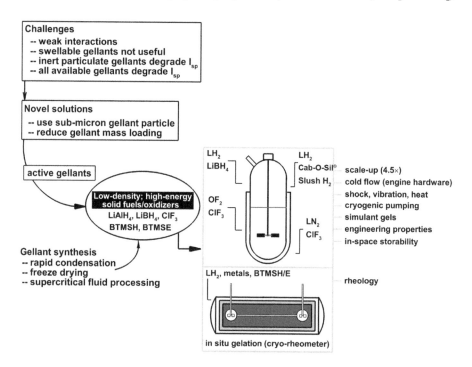

FIGURE 9.1 Development protocol for cryogenic gel propellants [6–8].

are low-density (70 kg/m^3 for LH$_2$), high-energy chemical compounds liquefied at very low temperatures (normal boiling point, T_{nb} = 20.3 K for LH$_2$).

They do not interact (i.e. do not form bonds or loose associations) with ordinary gellant molecules at such extreme conditions. These weak interactions mean that the fuel is not 'taken up' by the gellant and most organic gellants cannot be used for cryogenic propellants. It was established early on that particulate gellants could trap the cryogens [5]. Commercial silica-based particulate gellants (for example, Cab-O-Sil®) are readily available, though the low-density of cryogens necessitated high mass loading or nanometric particle size of these inert gellants for gaining appreciable up-take of the cryogen. However, chemical inertness of silica severely degrades specific impulse. A novel solution was therefore devised in the form of low-density, high-energy sub-micron particles of solid fuels to advance the development of cryogenic gel propellants. These gellants included aluminum (Al) and boron (B) metal hydrides of lithium (Li) fuel [4] and chlorine trifluoride (ClF$_3$) oxidizer [6]. Acid polymerization of 1,2-bis(trimethoxysilyl) ethane (BTMSE) and 1,2-bis(trimethoxysilyl)hexane (BTMSH) gave polymeric gellants [7, 8]. All these gellants are amenable to processing methods like rapid condensation, freeze-drying, and supercritical fluid processing that yield sub-micron gellant particles. Meanwhile, the dramatic decrease in the mass loading of gellant achieved using Cab-O-Sil® particles of size ~7 nm has been demonstrated along with 'slush' H$_2$ produced through ultra-low temperature cooling of LH$_2$ [5]. Slush H$_2$ is a mixture of liquid and solid H$_2$. In fact, all examples of cryogenic gel propellants mentioned have optimized on the concentration of gellant by bringing down the mass loading of gellant to less than 5 wt % through the use of sub-micron gellant particles.

The processing of cryogenic gel propellants adheres to all requirements of low-temperature handling, heat transfer, sealing, and material compatibility; thus leading to considerably complex gel processing facilities. Readers should consult detailed reports [4–6] for more discussion on this issue.

9.2.2 Gelation of Hypergolic Propellants

Hypergolic liquid propellants constitute another important class of highly energetic liquid propellants [3], which enable simpler combustion systems. Unfortunately, hypergols are also some of the most hazardous propellants that require very careful handling. Consequently, they also have been subjected to many experiments aimed at reducing their handling hazards while maintaining their hypergolic nature.

Hypergolic liquid propellant systems include pairs of fuel and oxidizer like unsymmetrical dimethyl hydrazine ((CH$_3$)$_2$N$_2$H$_2$, UDMH) and red fuming nitric acid (HNO$_3$, RFNA) and many other hypergolic combinations are known [3]. All these bipropellant combinations could be gelled separately and used, with a wide range of gellants to choose from [9]. Easy mixing is another feature of the development of hypergolic gel propellant. The requirements during mixing are not as stringent as cryogenic gel propellants though all essential handling and storage protocols for hypergolic liquid propellants must be strictly followed [10].

Laboratory-scale development of gel propellants has been attempted almost exclusively such that most of the work on gel propellant development reported so far is within the range of a few tens of grams to a few kilograms. However, important lessons have been learnt during large-scale development [11, 12].

9.2.3 NEW DEVELOPMENTS

Advances in the development of gel propellants have been made in many areas, some of which have seen benefits derived from advances in liquid propellants while some are specifically devised for gel propellants. Among them, improved mixing technology [13, 14], use of safer hypergolic propellant like dimethyl ethanamine azide (DMAZ) [13], and sodium borohydride ($NaBH_4$) additive for inducing hyper-golicity in non-hypergols [15] are some of the encouraging developments.

Conventional mixing is adequate in many propellant development programs, especially when gelation occurs upon storage. However, in some cases, rapid increase in viscosity occurs during mixing. Such cases have been handled using ultrasonic mixing in the past, and more recently, using low-frequency, high-intensity resonant acoustic mixing [13, 14]. Resonant oscillations apparently induce high dispersion by turbulence [13, 14], resulting in shorter mixing time and higher viscosity in comparison to the conventionally mixed gel propellant. The correct mixing time and acceleration for resonant acoustic mixing must be found after trials.

Comprehensive development of gel propellant [16] shows how the protocols encompassing connections among formulation, rheology, and atomization of gel propellants could be developed. Such a protocol is capable of yielding an optimal formulation of gel propellant that balances the conflicting demands of a sufficiently stable and yet atomizable gel propellant.

9.3 FLOW CHARACTERIZATION

The primary variable affecting the flow of gel is shear rate. Shear is imposed by the wall when gel flows through a channel and by the ambient gas phase when the gel flows in the forms of a free jet, sheet, and droplets in the combustion chamber. Extensional shear (longitudinal stretching) is also imposed on the gel during flow. Since gel propellant behaves like a solid during storage, its flow characterization also encompasses the transition from a static solid state. The consideration of limiting conditions approaching the static solid state is very important. In the discussion that follows, our emphasis is on interpretation of the flow characteristics in relation to the other elements of gel propellant technology.

9.3.1 RHEOLOGICAL PROPERTIES

In this section, important properties that characterize the flow of a non-Newtonian fluid such as gel propellant are briefly described and their importance for the performance of gel propellant is considered. As the name suggests, non-Newtonian fluids exhibit a non-linear relationship between the applied shear rate ($\dot{\gamma}$) and the induced shear stress (τ_{shear}) as shown in Figure 9.2a. The simplest relationship is described by the power law, $\tau_{shear} \propto \dot{\gamma}^n$, although more complicated behaviors are known [17].

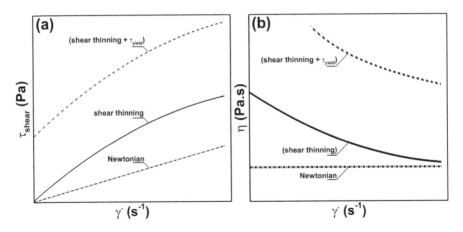

FIGURE 9.2 Non-Newtonian characteristics of gel propellants depicted in terms of shear thinning and yield stress: (a) shear stress-shear rate and (b) shear viscosity-shear rate curves.

Among the many types of non-Newtonians, shear-thinning fluids are most relevant for the discussion of gel propellants. As shown in Figure 9.2b, such fluids lose their viscous resistance ('thinning') upon increasing the shear so that their shear viscosity (η) approaches that of a much less viscous Newtonian fluid. The flow of a shear-thinning fluid could be complicated by the existence of a yield stress (τ_{yield}), also shown in Figure 9.2a. Presence of yield stress, the minimum shear stress above which a non-Newtonian fluid begins to flow, complicates the response of a shear-thinning fluid and delays the subsequent processes.

Apart from these easily visualized properties, non-Newtonian fluids also exhibit varying degrees of viscoelastic behavior characterized by elastic elongation along with a typically viscous response when subjected to shear and elongational stretch. The elongational response is expressed in analogous terms to the shear-thinning response using elongational viscosity and stretch rates. Many non-Newtonian fluids are also thixotropic (i.e. their response to shear is time-dependent). Thixotropic nature could be expressed in all properties of a non-Newtonian fluid. It should be noted that none of these properties are mutually exclusive; often all of these characteristics are present simultaneously in non-Newtonian fluids, and gel propellants are not an exception to this observation.

Measurements of these properties along with a comprehensive discussion can be found in specialized texts [17]. However, a glimpse of these rheological properties at work can be obtained and their relationship with gel propellant development can be illustrated by considering the effects of variables used in propellant processing. We undertake this exercise in the next section.

9.3.2 RELATIONSHIP BETWEEN COMPOSITION AND RHEOLOGY

Composition of gel propellant profoundly affects its rheological properties. The exact mechanisms are varied and rooted in the microstructure of gel propellant. We now consider selected results from literature to summarize the possibilities. Gellant

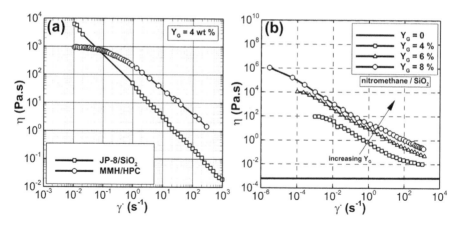

FIGURE 9.3 Effects of (a) gellant type [18] and (b) gellant concentration [19] on shear viscosity.

type, gellant concentration, mixing method, and mixing temperature are included as relevant parameters for gel propellant processing in this discussion and their effects on shear-thinning behavior, magnitude of yield stress, thixotropic nature, and viscoelasticity of gel propellants are considered. The measurements were conducted using rheometers.

Figure 9.3 shows that gel propellants of JP-8—monomethyl hydrazine ($CH_3N_2H_3$, MMH), and nitromethane (CH_3NO_2, NM) derived from silica (SiO_2) and hydroxy propyl cellulose (HPC) gellants—exhibit shear thinning [18]. The exact behavior depends on the gellant; HPC is a polymeric gellant and SiO_2 is an inorganic particulate gellant. For the same concentration of gellant (Y_G), shear viscosity of JP-8/SiO_2 gel propellant decreases continuously in comparison to MMH/HPC gel propellant, which also shows a plateau region where the viscosity is nearly independent of applied shear, as shown in Figure 9.3a.

The differences in the behavior of shear-thinning gels are due to different microstructures; JP-8/SiO_2 gel forms by joining the fuel and gellant molecules by H-bonds, and MMH/HPC gel contains a more complex structure induced by polymeric gellant. Moreover, thixotropic nature and yield stress of JP-8/SiO_2 gel also affect its shear-thinning characteristic [18]. Increase in gellant concentration increases the shear viscosity, as illustrated in Figure 9.3b.

However, for NM/SiO_2 gel propellant, higher Y_G also increases the degree of shear thinning, as reflected in the decreased shear viscosity at high shear rates. The increased degree of shear thinning irrespective of Y_G at higher shear levels has been described as a consequence of hydrodynamic forces overwhelmingly dominating the cohesive particle-particle interactions [19].

Low-frequency, high-intensity resonant acoustic mixing at 61.3 Hz requires ~240 s to mix 4–6 wt % SiO_2 in JP-8 as against ~4000 s required by conventional impeller blade rotating at 600 rpm [14]. Better mixing of JP-8/SiO_2 suspensions has been shown to produce a gel propellant of higher viscosity at low shear rates [14] as observed from Figure 9.4a, leading to a potentially better stability and longer useful life.

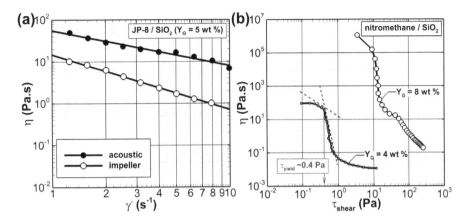

FIGURE 9.4   Influence of (a) mixing method on shear viscosity [14] and (b) gellant concentration on yield stress [19].

Figure 9.4b illustrates the effects of gellant concentration on the yield stress (τ_{yield}) of NM/SiO$_2$ gel propellant and demonstrates how τ_{yield} could be estimated by the tangent method. It indicates that particle-particle interactions at low shear are overtaken by hydrodynamic interactions at the critical shear stress (τ_{yield}) and irreversible flow of gel commences. For stresses smaller than τ_{yield}, these interactions enable the gel to behave like an elastic solid [19]. In Figure 9.4b, stronger interactions increase the τ_{yield} by an order of magnitude as the gellant concentration is increased from 4 to 8 wt %.

Shear thinning, as clear from the preceding discussion, pertains to a decrease in shear viscosity in response to shear because of the diminishing strength of the gel microstructure. The shear-thinning response of some gel propellants evolves over time once the shear is applied. Some gels even tend to recover gradually once the shear is removed. The temporal dependence of response as depicted in Figure 9.5 is one measure of thixotropy.

The extent of thixotropic nature is explained by various mechanisms and it is also affected by gellant concentration, as shown in Figures 9.5a and 9.5b for JP-8/SiO$_2$ gel propellant [20]. A constant shear rate ~10^3 s^{-1} was applied and then removed to obtain the response shown in Figure 9.5a. A recovery period of ~480 s followed by another application of the same magnitude of shear showed the response as in Figure 9.5b. It is observed that a higher degree of shear thinning induced for $Y_G = 7$ wt % also made the gel less thixotropic since the response time shortened to ~25 s from ~150 s. The lack of recovery after rest (Figure 9.5b) suggests that these gels are 'irreversibly' thixotropic [20].

Viscoelastic behavior is characterized by storage modulus (G') and loss (shear) modulus (G''). Larger values of these moduli represent dominant behavior of elastic and viscous branches, respectively. Gel for which $G' \gg G''$, expresses dominantly elastic behavior, and vice versa. The dynamic response of the viscous and elastic features of gel propellant is illustrated in Figure 9.6. The rheometer was operated at different angular frequencies.

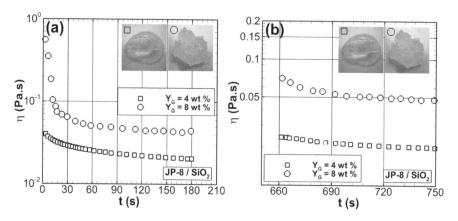

FIGURE 9.5 Influence of gellant concentration on the irreversible thixotropic behavior [20, 21], (a) $\dot{\gamma} = 10^3\ \text{s}^{-1}$ applied to convert gel into sol and (b) $\dot{\gamma} = 10^3\ \text{s}^{-1}$ applied after the resting period of ~480 s.

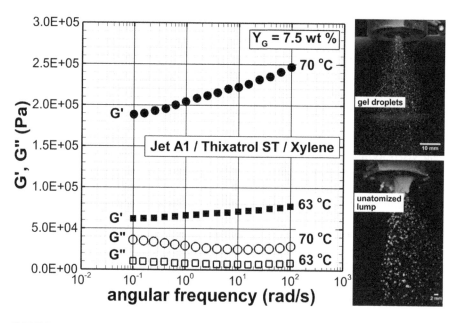

FIGURE 9.6 Influence of mixing temperature on storage and shear moduli as indicators of viscoelastic behavior of gel propellant and quality of atomization [22].

Smaller frequencies represent the long-duration events such as storage, while higher frequencies simulate the fast processes during breakup of sheets and films. High storage modulus is favorable for storage. On the other hand, a high shear modulus and frequency-independent response of the moduli are desirable traits for atomization. This is because a higher storage modulus tends to delay the breakup and impair atomization. Another issue of importance is the dependence of response on frequency. A frequency-dependent response that enhances storage modulus at higher frequencies (fast events like atomization) also impedes atomization. This feature is illustrated for Jet A1/Thixatrol ST/Xylene gel propellant in Figure 9.6 with the help of mixing temperature as a parameter for gel propellant development [22]. Thus, a propellant processed at lower temperature (inset image at top) is easily atomized in comparison to a higher processing temperature (inset image at bottom). Clearly, formulation parameters should be given careful attention during the development of gel propellant to strike a balance between storage and atomization properties.

9.4 ATOMIZATION OF GEL PROPELLANTS

Since all gel propellants must be atomized like any conventional liquid propellants, the breakup of the bulk state of gel propellants has been extensively investigated [2]. The interactions among formulation, rheology, and atomization play a critical role in the success of atomization technology adopted for gel propulsion or developed specifically for the purpose. In this chapter, we get a glimpse of these interactions with the help of specific results and explain some of the challenges for successful atomization.

9.4.1 IMPINGING JETS ATOMIZATION

Impinging jet atomization is a commonly employed method for rocket engines [3]. It consists of two or more inclined jets of propellant impinging on each other to produce a thin unstable sheet that eventually breaks up into ligaments and droplets. The doublet, triplet, and quintuplet configurations of propellant jets are composed of same propellant (fuel or oxidizer) or dissimilar propellants (fuel and oxidizer).

Impinging jet atomization of gel propellants has been achieved and equivalent performance in comparison with liquid propellants is already demonstrated [2]. This success is achieved by exploiting the shear-thinning behavior of gel propellants using sufficiently high jet velocities in the nozzle. The breakup of gel propellants could be analyzed using regime maps, as illustrated in Figure 9.7.

The breakup regimes identified using visualization of sprays are plotted in terms of dimensionless groups generalized Reynolds number (Re_{gen}) and jet Weber number (We_j) as well as processing parameters such as concentration of nano particles (Y_A). All practical engines must operate in the impact wave (●, ○) and prompt breakup (×) regimes because the droplet sizes are sufficiently small and well distributed for a stable, self-sustained combustion.

Typical distributions of the droplet populations are shown in Figure 9.8 for the same gel propellant as in Figure 9.7. They are shown here in terms of the Sauter mean diameter (D_{32}) for the impact wave regime (●, ○) at a supply pressure (P_S) of 828 kPa. D_{32} is useful to characterize the combustion systems relying on volumetric heat release and vaporization of droplet. Droplet diameter and length coordinates are non-dimensionalized using the orifice diameter (d_o) of the injector.

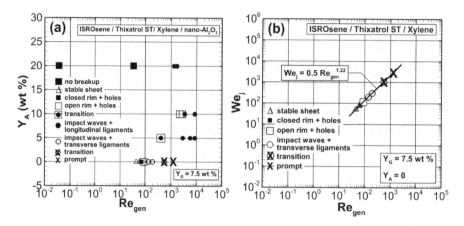

FIGURE 9.7 Breakup regimes observed in doublet impinging jet atomization of gel propellant as a function of (a) nanoparticle concentration and generalized Reynolds number and (b) jet Weber number and generalized Reynolds number [23].

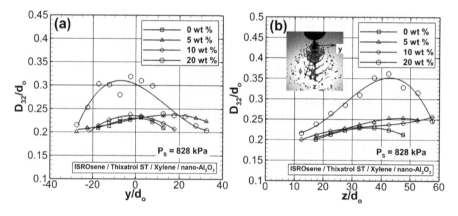

FIGURE 9.8 Distributions of mean droplet diameters of gel propellant sprays formed using doublet impinging jets in the impact wave regime; (a) y direction and (b) z direction [24].

Figure 9.7 reveals that the most useful impact wave and prompt breakup regimes are achieved only at progressively more severe operating conditions when a gel propellant of higher nano particle loading is used. This is a direct consequence of higher shear viscosity of such highly loaded gel propellants during storage. In this sense, gel propellants are potentially more difficult to atomize than liquid propellants. Moreover, although the droplets of gel propellants produced by impinging jets atomization are similar to liquid propellants, coarser droplets are produced for highly loaded gel propellant as shown in Figure 9.8. Coarser droplets carry more fuel mass but they also burn more slowly than finer droplets, which could negatively affect the flame stabilization.

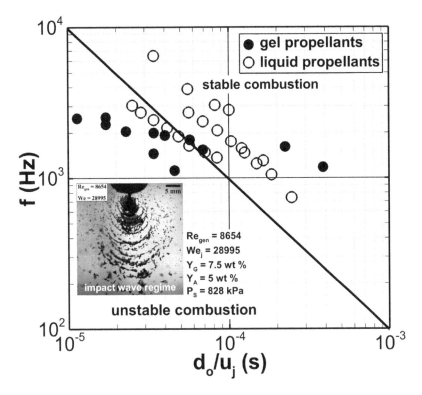

FIGURE 9.9 Possibility of combustion instability induced by gel propellants in comparison with conventional liquid propellants [24].

Further, the increased viscous resistance of gel propellants loaded with higher amounts of nano particles must be overcome by employing higher shear rates in the nozzle and/or higher jet velocities (u_j) so that the impact wave regime of sheet breakup is reached. As the name suggests, breakup of the sheet of gel propellant in this regime is governed by periodic surface waves emanating from the point of impact of the two jets, as shown in the inset of Figure 9.9.

The frequency of impact waves (f) is plotted in Figure 9.9 against the combined influence of d_o and u_j in terms of the parameter (d_o/u_j). We observe that higher shear viscosity of gel propellants restricts the injector parameter (d_o/u_j) to smaller values. Impact wave frequencies arising due to restrictions on d_o (as small as possible) and u_j (as large as possible) could lead to combustion instability. On the other hand, negligible viscous resistance of liquid propellants simplifies the choice of (d_o/u_j) to avoid combustion instability.

Moreover, provision of high shear rate in the orifice of gel propellant is in itself not always adequate to produce a propellant spray of desired properties. This possibility is illustrated for gel propellant using the data presented in Figure 9.9. A water-Carbopol gel propellant simulant was developed with shear-thinning behavior quantitatively similar to the actual gel propellant as shown in Figure 9.10a.

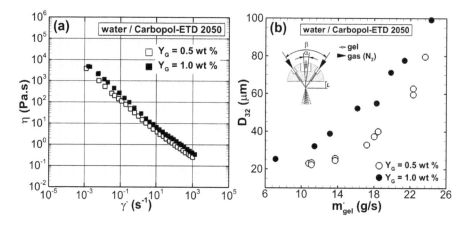

FIGURE 9.10 Performance of gel propellant simulants on conventional triplet impinging jets: (a) shear-thinning behavior [25] and (b) variation of mean droplet diameters in response to change in the mass flow rate of gel [25].

The gel was subsequently atomized using a triplet impinging jets atomizer by transforming the jet of a gel stream into a sheet upon oblique impingement by two gas jets as shown in the inset of Figure 9.10b. Thus, the two-fluid atomizer used kinetic energy of atomizing gas to destabilize the jet of gel stream. Measurements of droplet sizes acquired for a fixed set of parameters α, β, L, and the flow rate of atomizing gas (N_2) are shown in Figure 9.10b for two gellant concentrations and a range of mass flow rates of the gel ($\dot{m}_{gel}$).

We observe that even though the higher shear caused by increased flow rate of gel would have reduced the viscous resistance of gel, increased mass would be forced into the sheet for a fixed flow rate of atomizing gas. The resulting kinetic energy would be insufficient to produce a thin sheet. This would cause ligaments and droplets of a larger diameter.

9.4.2 NEW ATOMIZATION METHODS

In atomization, the goal is to obtain the desired quality of spray in terms of droplet size and size distribution. In this respect, the need to achieve atomization using less severe operating conditions is clearly indicated from the preceding discussion. Atomization must also be reliable when the flow rate of gel propellant is required to be changed according to the thrust program. In literature, many alternative methods for atomization of gel propellants have been proposed. In this section, we would consider one novel method that utilizes the internally impinging jets [26].

As shown in Figure 9.11a, the internally impinging atomizer [26] includes an impingement zone created at the bottom end (shown highlighted in Figure 9.11a) with the help of eight microjets of air (jet diameter 0.4 mm). These microjets impinge on each other and help to form a system of multiple confined vortices. This flow field generates shear on the exposed surface of the gel propellant stream and induces

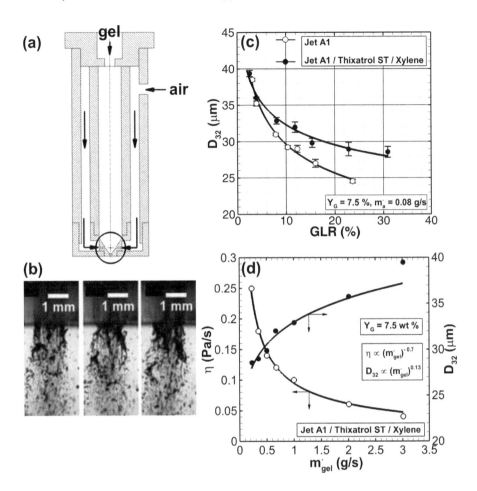

FIGURE 9.11 An example of efficient two-fluid atomization of gel propellant; (a) internally impinging atomizer [26], (b) snapshots of breakup of the gel propellant film by atomizing air ($\Delta t = 59$ µs) [26], (c) influence of shear viscosity on mean droplet diameter [26], and (d) influence of flow rate of gel propellant on shear viscosity and mean droplet diameter [22].

shear thinning. The gel stream turns into a thin film that flows downstream along the walls of the conical section and nozzle. It partly breaks up within the atomizer. The partially broken up film produces longitudinally aligned ligaments along with droplets, as shown in Figure 9.11b. Remaining breakup processes are completed as shown in the snapshots some distance downstream of the nozzle exit plane.

An important feature of the internally impinging atomizer is its ability to produce droplets well within the typical requirement for a ramjet engine (D_{32} ~50 µm) as observed from Figure 9.11c. Moreover, the same atomizer hardware is equally useful for a liquid propellant and its gel propellant. This atomizer has also been employed in spray combustion. While the effect of viscous resistance is manifested in Figure 9.11c, we also observe in Figure 9.11d that the arguments associated with

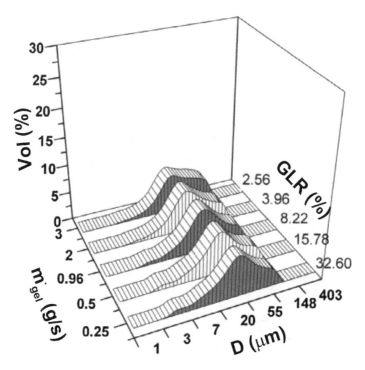

FIGURE 9.12   Performance of novel internally impinging atomizer in terms of volume distributions of droplets as a function of mass flow rates of gel propellant [26].

increased film/sheet thickness with increase in the mass of gel propellant (at constant mass flow rate of atomizing air, $\dot{m}_a = 0.08$ g/s) lead to relatively coarser droplets, even though viscosity is reduced by higher shear levels. In this manner, features of the triplet impinging jets atomizer [25] and the internally impinging atomizer are qualitatively similar.

Figure 9.12 suggests that increasing the mass flow rate of gel propellant actually increases the volume fraction of coarser droplets in the sample population, reduces the fraction of fine droplets, and produces broader distributions.

9.5   SPRAY COMBUSTION OF GEL PROPELLANTS

In this section, we consider the basic features of combustion of a gel propellant in the form of a spray. In spray combustion, the processes occurring at the droplet level influence the combustion of an ensemble of droplets. Spray combustion is an important phase that when successful, always indicates that the given gel propellant was formulated with adequate care, possesses requisite rheological properties, and had been atomized with a good quality of spray in terms of droplet size and distributions. To date, few gel propellants have been taken through these steps and have achieved successful combustion in laboratory and actual applications. We consider one case to illustrate some unique features of gel propellant combustion and also highlight inherent complexities.

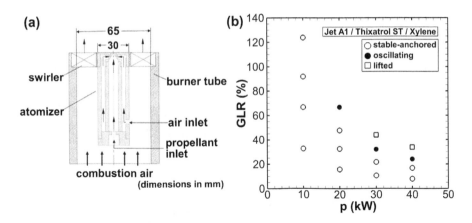

FIGURE 9.13 (a) Swirl burner fitted with the internally impinging atomizer as a bluff body and (b) regime map of gel propellant spray flames stabilized on the burner (Y_G = 7.5 wt %) [27].

An optimal composition of Jet A1 gel propellant has been developed [16], shown to have balanced rheological properties [22], and has been atomized using the internally impinging atomizer [26]. The balance of rheological properties has been found by exploiting the interactions among formulation, rheology, and atomization [22, 26]. Internally impinging atomizer was then integrated with a laboratory swirl burner as shown in Figure 9.13a [27]. In addition to its main role, the atomizer also acts as a bluff body and helps in the mixing and flame stabilization. A large number of experiments performed on this integrated atomizer-burner system have been reported [27]. The overall behavior of spray flames of Jet A1-Thixatrol ST-Xylene gel propellant and swirling combustion air (Swirl number S_N = 0.9) is summarized in Figure 9.13b.

The spray flame could be classified into three highly reproducible regimes controlled by the ratio by mass of atomizing gas and propellant (*GLR*) used for atomization and nominal flame power (*p*), the equivalent thermal energy obtained by complete combustion of fuel at a specified flow rate. These regimes indicate that the previous developmental phases have been successful and further suggest that the spray flames of gel propellant could behave similarly to liquid propellant flames in some respect. Further illustration of their morphological features is given in Figure 9.14. The images of natural luminosity shown in Figures 9.14a–9.14d are for a 20 kW flame under the same conditions as in Figure 9.13b. Increase in the *GLR* (achieved by the flow rate of atomizing air $\dot{m}_a$) gradually transforms the flame into a state of oscillating flame for which the stabilization plane is frequently displaced in vertically upward and downward directions with a well-mixed flame base (Figure 9.14d). The flame core shown by high-speed backlit visualization in Figure 9.14e is composed of an evaporating core of gel propellant spray surrounded by the corrugated flame front for a condition corresponding to Figure 9.14a. Another view of the natural luminosity is shown in Figure 9.14f.

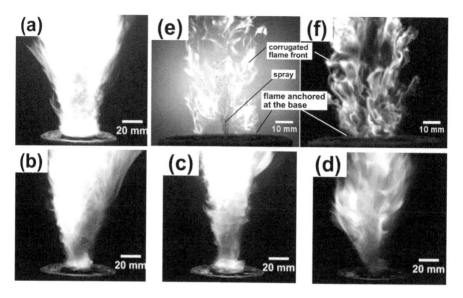

FIGURE 9.14 Morphological features of the spray flames of Jet A1-Thixatrol ST-Xylene gel propellant [27]; (a–d) anchored and lifted flames for 20 kW power, (e) backlit image of the stable anchored spray flame showing inner core of spray, and (f) spray flame under same operating conditions as (e).

Experiments using the same atomizer-burner system and operating conditions have also been performed on Jet A1 propellant. They have helped to document the differences between liquid propellant and its gel propellant. Figure 9.15 shows that the flame of gel propellant spray is more compact, although the major heat release region is concentrated in the jet-like propagation zone for both propellants.

The flame neck and base regions do not yield much heat release in gel propellant. These observations are due to larger droplet sizes of the gel propellant, longer times for vaporization and combustion, and possible attenuation of turbulence by higher viscosity of gel droplets. However, other factors that tend to increase the combustion efficiency of gel propellant spray and cause a compact flame have been identified [27]. These include the occasionally intense combustion of gel droplets triggered by jetting events, increased oxygen levels due to gellant (leaner combustion) and thermal radiation. Overall behavior of gel propellant leads to a compact flame with visible height ~158±2 mm compared with ~300±3 mm of Jet A1 flame when flow is swirling [27]. The gel propellant induces differences because similar behavior is observed for non-swirling flames.

9.6 CONCLUDING REMARKS

Gel propellants are highly viscous, shear-thinning, non-Newtonian fuels and oxidizers of complex microstructure that enables them to be stored for indefinitely long duration and to be used like liquid propellants. These potentially beneficial characteristics of gels must be carefully investigated and modified to suit the needs

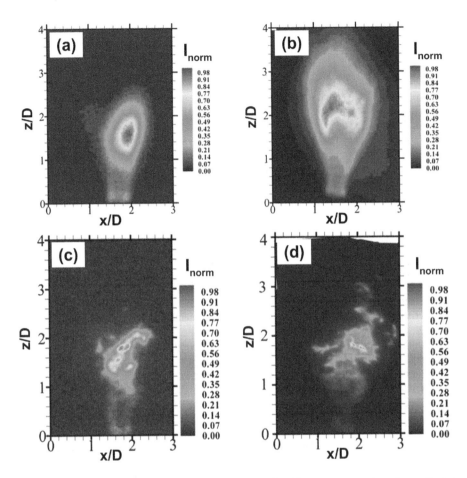

FIGURE 9.15 High-speed line-of-sight CH* chemiluminescence imaging of Jet A1-Thixatrol ST-Xylene gel propellant/air spray flame in the left panel and Jet A1 propellant/ air spray flame in the right panel [27]; (a) and (b) are average, (c) and (d) are instantaneous images.

of practical applications. Most of the current and future applications are related to rocket propulsion systems.

Gel propellants have come a long way since the beginning with impressive, if at times discontinuous, progress towards realization of flight-worthy gel propulsion systems. Propellant development, flow characterization, atomization, and combustion, the four basic elements of a gel propulsion system, have witnessed adequate progress to the extent that technology demonstrator flights have already been achieved in recent times.

The future progress in gel propulsion system depends largely on our ability to control and balance their rheological properties such that atomization and combustion could be accomplished in a high-pressure operation. Emphasis on any one of the elements is generally counter-productive and ultimately limits the realizable potential of gel propellants.

REFERENCES

[1] Calabro M. Overview on hybrid propulsion. *Prog Propul Phys* 2011;2:353–374.
[2] Natan B, Rahimi S. The status of gel propellants in year 2000. In: *Combustion of Energetic Materials*, Kuo KK and deLuca L (eds.), Begel House, Boca Raton, 2001.
[3] Mishra DP. Ch. 6 Chemical rocket propellants. In: *Fundamentals of Rocket Propulsion*, CRC Press, Boca Raton, Florida, USA, 2017, p. 161.
[4] McKinney CD, Tarpley WB. Gelling of liquid hydrogen. 1966;NASA-CR-54967:1–92.
[5] Rapial AS, Daney DE. Preparation and characterization of slush hydrogen and nitrogen gels. Technical Note 378, National Bureau of Standards, 1969.
[6] Globus RH, Beadle PD, Beegle RL, Cabeal JA. System analysis of gelled space-storable propellants. Summary Report 1038–02S, July 1968;NASA CR-100186:1–179.
[7] Wong W, Starkovich J, Adams S, Palaszewski B. Cryogenic gellant and fuel formulation for metallized gelled propellants: Hydrocarbons and hydrogen with aluminum. AIAA-94–3175. 30thAIAA/ASME/SAE/ASEE Joint Propulsion Conference, 27–29 June 1994, Indianapolis, IN, USA.
[8] Starkovich J, Adams S, Palaszewski B. Nanoparticulate gellants for metallized gelled liquid hydrogen with aluminum. AIAA-96–3234. 32nd Joint Propulsion Conference and Exhibit, 1–3 July 1996, Lake Buena Vista, FL, USA.
[9] Varghese TL, Gaindhar SC, John D, Josekutty J, Muthiah RM, Rao SS, Ninan KN, Krishnamurthy VN. Developmental studies on metallised UDMH and kerosene gels. *Def Sci J* 1995;45(1):25–30.
[10] Rahimi S, Hasan D, Peretz A. Development of laboratory-scale gel-propulsion technology. *J Propul Power* 2004;20(1):93–100.
[11] Pinns ML, Olson WT, Barnett HC, Breitwieser R. NACA research on slurry fuels. NACA Lewis Flight Propulsion Laboratory, Report 1388, 1958:1–28.
[12] Caves RM. Stabilization of 50-percent magnesium-JP-4 slurries with some aluminum soaps of C_8 acids. NACA Lewis Flight Propulsion Laboratory, NACA RM E54C10 1954:1–50.
[13] Coguill SL. Synthesis of highly loaded gelled propellants. Technical Report, Resodyn Corporation, Butte, MT, 2009.
[14] Arnold R, Santos PHS, Kubal T, Campanella O, Anderson WE. Investigation of gelled JP-8 and RP-1 fuels. Proceedings of the World Congress on Engineering and Computer Science (WCES 2009), 20–22 October 2009, San Francisco, USA.
[15] Connell Jr TL, Risha GA, Yetter RA, Natan B. Ignition of hydrogen peroxide with gel hydrocarbon fuels. *J Propul Power* 2018;34(1):170–181.
[16] Padwal MB, Mishra DP. Synthesis of Jet A1 gel fuel and its characterization for propulsion applications. *Fuel Process Technol* 2013;106:359–365.
[17] Chhabra RP, Richardson JF. Rheometry for non-Newtonian fluids. In: *Non-Newtonian Flow and Applied Rheology*, Butterworth-Heinemann, Burlington, USA, 2008.
[18] Arnold R, Santos PHS, deRidder M, Campanella OH, Anderson WE. Comparison of monomethylhydrazine/hydroxypropylcellulose and hydrocarbon/silica gels. AIAA-2010–422. 48th AIAA Aerospace Sciences Meeting including the New Horizons Forum and Aerospace Exposition, 4–7 January 2010, Orlando, Florida, USA.
[19] Teipel U, Förter-Barth U. Rheological behavior of nitromethane gelled with nanoparticles. *J Propul Power* 2005;21(1):40–43.
[20] Santos PHS, Arnold R, Anderson WE, Carignano MA, Campanella OH. Characterization of JP-8/SiO2 and RP-1/SiO2 gels. *Engineering Letters* 2010;18:1–9.
[21] Santos PHS, Carignano MA, Campanella OH. Qualitative study of thixotropy in gelled hydrocarbon fuels. *Engineering Letters* 2011;19:1–7.

[22] Padwal MB, Mishra DP. Interactions among synthesis, rheology, and atomization of a gelled propellant. *Rheol Acta* 2016;55:177–186.

[23] Jejurkar SY, Yadav G, Mishra DP. Visualizations of sheet breakup of non-Newtonian gels loaded with nanoparticles. *Int J Multiph Flow* 2018;100:57–76.

[24] Jejurkar SY, Yadav G, Mishra DP. Characterization of impinging jet sprays of gelled propellants loaded with nanoparticles in the impact wave regime. *Fuel* 2018;228:10–22.

[25] Rahimi S, Natan B. Atomization of gel propellants through an air-blast triplet atomizer. *Atomization Spray* 2006;16(4):379–400.

[26] Padwal MB, Mishra DP. Characteristics of gelled Jet A1 sprays formed by internal impingement of micro air jets. *Fuel* 2016;185:599–611.

[27] Padwal MB, Mishra DP. Experimental characterization of gelled Jet A1 spray flames. *Flow, Turbul Combust* 2016;97:295–337.

10 Developmental Study of Aluminized Fuel-Rich Propellant

Nikunj Rathi and P.A. Ramakrishna

CONTENTS

10.1 NOMENCLATURE

v_o = vehicle flight velocity, m/s
ε = air-fuel ratio (A/F)
I_{sp_CEA} = specific impulse from NASA SP273, m/s
I_{sp_ram} = ramjet specific impulse, s

10.2 INTRODUCTION

Aerospace propulsion can be broadly classified into air breathing and non-air breathing propulsion. Rockets fall under the category of non-air breathing propulsion. The advantage that an air breathing propulsion system has over a non-air breathing system is that the former carries no oxidizer to produce the thrust, and therefore delivers very high specific impulse (I_{sp}). However, the air breathing propulsion has its own limitations, one among them is the top speed achieved. There has been a concerted effort worldwide to increase the speed envelope of an air breathing engine in order to obtain higher I_{sp} for a system as a whole. A ramjet is an air breathing propulsion system which can attain a supersonic Mach number as high as 3.5. They have been in use for a long time in missiles due to their high I_{sp} compared to rocket engines.

A ramjet has no moving parts like turbines and compressors and therefore it cannot produce any static thrust. It needs a booster system which can take it to supersonic speeds, beyond which it starts producing net positive thrust. The incoming

DOI:10.1201/9781003049005-10

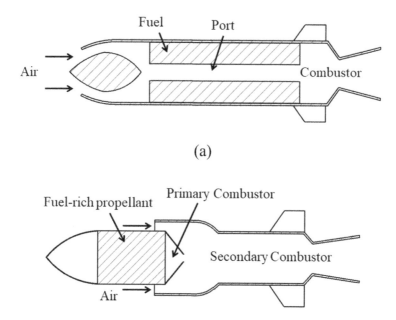

FIGURE 10.1 Schematic of SFRJ in (a) hybrid mode and (b) two combustor chamber.

supersonic air is compressed through an oblique shock and finally a normal shock bringing the flow to subsonic speeds at the entrance of the combustion chamber. The compressed subsonic air then reacts with the fuel in the combustion chamber resulting in high temperature and high pressure gases, which are expanded through a convergent divergent nozzle to produce thrust.

Typically, ramjets are liquid fuelled and use kerosene as fuel and air as oxidizer. Being a liquid fuel, it has to atomize, vaporize, mix and then react with the incoming compressed air. This increases the residence time required or the length of the combustion chamber for complete reaction. Also, kerosene has a density of 800 kg/m³. On the contrary, a solid fuel ramjet (SFRJ) utilizes solid fuel-rich propellant, which has a higher density of around 1500 kg/m³, making the overall system compact.

There are two types of SFRJ configurations that have been described in literature—namely, the hybrid mode and the dual combustion chamber mode. The hybrid mode consists of a port burning fuel grain with air passing through the fuel port as shown in Figure 10.1a. The solid propellant required to provide the initial boost to take the vehicle to the required Mach number for the ramjet to start functioning has to be housed separately.

In the dual combustion chamber mode, a fuel-rich propellant (FRP) burns in the primary combustor and ejects fuel-rich hot gases through a choked nozzle as shown in Figure 10.1b, which burns completely with air in the secondary combustor and provides thrust. The secondary combustor houses the solid propellant which provides the initial boost to take the vehicle to required Mach number for the ramjet to start functioning.

10.3   LITERATURE SURVEY

For the hybrid mode SFRJ, Krishnan and George (1998) have provided an excellent review of the various fuels used and work done by different countries in this area. The prominent ones that were not included in this article were Schulte (1986), Gobbo-Ferreira et al. (1999) and Raghunandan et al. (1985). Schulte (1986) and Gobbo-Ferreira et al. (1999) have used polyethylene and Raghunandan et al. (1985) have used polyester as fuel. The experiments were carried using both electrically heated air and vitiated air. Variation of burn rate on air mass flux, chamber pressure, inlet air temperature and rearward step height; and dependence of combustion efficiency on combustion chamber length were studied. The flammability limit as a function of port to throat area and equivalence ratio were also studied.

Among the two SFRJ configurations discussed, the combustion in the hybrid mode is boundary layer controlled, where the fuel diffuses into the air stream passing through the fuel port. This results in improper mixing and lower combustion efficiency as reported by Krishnan and George (1998). Besides, most times the fuels used are polymeric fuels which have lower density. In hybrid mode, it would not be possible to control the fuel flow rate on command as it is primarily controlled only by the air mass flux. On the other hand, the dual combustor ramjet has a fuel-rich propellant, which is often metallized making it more energetic and dense. The fuel could be injected through multiple throats into the secondary chamber as high temperature fuel-rich gases, which results in better combustion efficiency. The fuel flow rates could be changed by altering the throat area of the primary combustion chamber. Another aspect is that a separate solid propellant booster would be required for the hybrid mode operation, which makes it bulky. Owing to these shortcomings of the hybrid mode SFRJ, this study focuses on solid fuel ramjet with two combustion chambers.

The dual combustion chamber configuration for SFRJ makes use of a fuel-rich propellant. The fuels considered in the literature for this configuration are, broadly, glycidyl azide polymer (GAP) and AP/metal based. Kubota and Kuwahara (1991) have developed an energetic modified GAP based fuel. It was found to burn very rapidly even though the combustion temperatures were lower than conventional solid propellant. Two types of combustors were used by them, direct connect flow (DCF) and semi-freejet (SFJ). They report that the propellant showed a dual burn rate pressure index (n) behaviour. Combustion efficiency as a function of characteristic length reached 92% for 3 m length of secondary combustor. A peak specific impulse of 780 sec for an air-fuel ratio of 14 was reported. Kubota et al. (1991) used a throattable nozzle in the primary chamber to control the fuel flow rate. Modified GAP fuel without any metal fuel was compared with typical AP/HTPB and NC/ NG (nitrocellulose/nitroglycerin) based propellants. It was found that GAP modified fuel resulted in the highest theoretical specific impulse and burn rates compared to AP/HTPB and NC/NG based fuel.

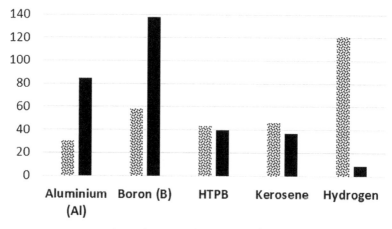

FIGURE 10.2   Heat of oxidation for different fuels.

In case of AP/metal based fuels, the idea of adding metal into the fuel arises from the fact that metals have been known to have higher equilibrium combustion temperatures than a hydrocarbon based fuel as shown by Kubota et al. (1992). Along with higher equilibrium temperature there is an added advantage of higher density with metals. Figure 10.2 is compiled based on inputs from Risha et al. (2007) and Gany (2006). Here, it compares the heat of oxidation of metals like Al and B with HTPB, kerosene and hydrogen. Among all fuels hydrogen has the highest heat release by weight. However, due to its very low density, the volumetric heat release is lowest among all fuels. Lower heat of combustion per unit volume means a lesser amount of fuel can be stored in a given volume. This is a pertinent issue as most aerospace systems would be limited by volume and as a consequence, density of the fuel plays a very crucial role.

Boron based metallized AP/CTPB composite propellants were studied by Kubota et al. (1992) to determine the combustion efficiency of boron particles in a direct connect flow combustor with a single-port and multi-port air inlet. Three different boron percentages were used, with 40% leading to the highest theoretical specific impulse and flame temperature. Propellants with 30% boron loading were made by them and the smallest boron particle size (2.7 μm) resulted in the highest burn rates and a pressure index (n) of 0.61. The combustion efficiency as high as 91% was reported with a multi-port air inlet compared to 79% with single-port.

Gany and Netzer (1986) studied the combustion of a boron based highly metallized solid fuel for SFRJ in a two-dimensional combustor using high speed photography. They found that usually the metal combustion did not happen at the propellant surface and it was emitted in the gas stream in form of segments and pieces. These segments and pieces merged together to form a bigger particle and the metal combustion occurred away from the propellant surface. Later, theoretical studies were carried out by Natan and Gany (1993) to investigate the effects of various parameters like inlet air temperature, air-fuel ratio, aft-burner length, etc. on boron combustion efficiency and chamber temperatures for different boron loadings. At low inlet air mass flux or high pressure, when the flow velocity was low (more residence time) the combustion efficiency and gas temperature was found to be high. The same was observed with a longer aft-burner length too. Similar results were reported by Kubota et al. (1992). Also, it was found that there is optimal boron loading for which highest boron combustion efficiency and gas temperature would be achieved for a fixed air-fuel ratio, pressure and air mass flux.

Gany and Netzer (1985) have presented an exhaustive survey of various metal hydrides, metal-metal compounds and borides that improve the energetics of a fuel-rich propellant. Metal hydrides are an attractive choice due to presence of hydrogen which increases heat of combustion per unit mass over the metal itself. However, in terms of per unit volume it is much less than pure metal due to its low densities. Boron carbide (B_4C) is a cheaper source of boron and very close to pure boron in performance. They have shown that among all these energetic compounds AlB_{12} has the highest energy density, even higher than the pure boron. Although some of these compounds have higher volumetric and gravimetric heat release than pure metals they were not used in the present study, as making these compounds requires an additional process. Also, these are not as easily available as pure metals.

Zhongqin et al. (1986) used an aluminum (20% by weight) based metallized solid fuel (AP (50%)/HTPB (17%) by weight) to study combustion efficiency in a SFRJ. With a single jet nozzle for the primary chamber, the combustion efficiency obtained by them was 57%, for a secondary combustion chamber length of 1100 mm. A maximum combustion efficiency of 77% was reported for a multiple jet nozzle. Experiments were carried out using different primary nozzles with multiple injectors and it was found that with the use of multiple primary nozzles the combustion efficiency increased by 30% owing to the enhanced mixing of primary hot gases with secondary airflow. Also, slag analysis at the secondary nozzle showed that with multiple jets the unburnt aluminum was reduced to zero from 48% with a single jet nozzle.

Recently, Nanda and Ramakrishna (2013) prepared fuel-rich solid propellants with Al, AP and HTPB. These propellants had energetics comparable to boron based fuel-rich propellant and were extremely cost effective. A variety of catalysts were used to achieve high burn rates. Along with that, different AP particle sizes were also employed. Recrystallization and double recrystallization techniques developed by Ishitha and Ramakrishna (2014a, 2014b) were also used on AP. The highest burn rate reported by Nanda and Ramakrishna (2013) was 3.8 mm/s at 70 bar for the propellant made with nano-sized iron oxide as catalyst. However, the minimum unburnt propellant achieved was of the order of 20–25%. The residue is defined as the percentage of the solid left unburnt (that does not undergo gasification and is left over in the solid phase itself) to the initial weight of the propellant sample.

10.4 MOTIVATION

As seen from the literature review, boron has been a preferred choice for the met-allized fuel-rich propellant due to its highest gravimetric and volumetric heat of oxidation as seen in Figure 10.2. However, it has certain disadvantages like a high melting and boiling point, ignition and combustion problems at higher boron content (Athawale et al. 1994). Therefore, to provide stable combustion, the boron content to be used in a fuel-rich propellant is limited. Also, from an economical point of view, it is around 300 times more expensive than aluminum (boron: 1300–1500 USD/kg and aluminum: 3.2–6.35 USD/kg) (Nanda and Ramakrishna 2013).

Apart from the economical point of view, the aluminum based fuel-rich propel-lant developed by Nanda and Ramakrishna (2013) had energetic properties simi-lar to typical boron based fuel-rich propellant developed by Kubota et al. (1992). Figure 10.3 shows the calculated ramjet specific impulse versus air-fuel (A/F) ratio for aluminum based fuel-rich propellant (Nanda and Ramakrishna 2013) and typical boron based fuel-rich propellant. The I_{sp} was obtained using CEA software (Gordon and McBride 1971), which assumes that air is carried on-board the system as an oxidizer. However, a ramjet utilizes atmospheric air as an oxidizer and hence is not carried on-board. Besides, the air comes into the ramjet engine with a certain momentum, which needs to be considered. Accounting for these two aspects, the expression for the theoretical specific impulse for a ramjet can be written as

$$I_{sp\_ram} = \left[(1+\varepsilon) I_{sp\_CEA} - \varepsilon v_o \right] \tag{10.1}$$

where ε is air-fuel ratio and v_o is vehicle velocity or air flow velocity at inlet. These calculations consider combustor and nozzle efficiencies to be unity.

It can be seen from Figure 10.3 that boron based fuel-rich propellant delivers slightly higher I_{sp_ram} compared to aluminum based fuel-rich propellant developed by Nanda and Ramakrishna (2013) beyond an air to fuel ratio of 15. However, this increment in I_{sp_ram} comes at a much higher cost of boron based fuel-rich propellant as explained earlier. Volumetric heat release (refer to Figure 10.2) for aluminum is twice that of HTPB, which would mean that if aluminum is used, it could result in a high density fuel, leading to a compact propulsion system. Therefore, an alumi-num based fuel-rich propellant seems to be promising option for a ramjet in terms of I_{sp_ram}, volumetric heat release, density and cost. Thus, in the present study alu-minum is used as a metal fuel for fuel-rich propellant. However, the residue or the unburnt propellant with aluminum based fuel-rich propellant observed by Nanda and Ramakrishna (2013) was 20–25%, which is quite high and needs to be brought down significantly for it to be useful.

The residue can be possibly brought down by enhancing the reactivity of AP and/ or Al, which would primarily result in greater pyrolysis of ingredients, leading to lower residue due to heat release occurring very close to the burning surface (Kohga 2011). Ishitha and Ramakrishna (2014a) used an activated charcoal (AC) catalyst embedding on AP to increase the burn rates of composite solid propellants. They carried out detailed studies to understand how AC embedded on AP could result in as much as 50% higher burn rates as compared to the case when they are just

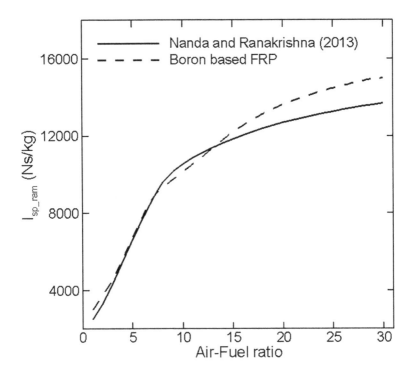

FIGURE 10.3   Calculated ramjet specific impulse versus air-fuel ratio.

mixed in a solid propellant with only 1% of catalyst addition. Marothiya et al. (2013) discussed the effectiveness of embedding iron oxide (IO) on AP. They report a 30% increase in burn rate with 1% IO using this technique as compared to just mixing. The fundamental reason why embedding catalyst (AC or IO) on AP was found to be more effective than mixing was that these catalysts act on AP alone (Ishitha and Ramakrishna 2014a; Marothiya et al. 2013). In the case of just mixing these catalysts in the propellant, it gets dispersed everywhere and the interaction sites between AP and catalyst are limited and are often hindered by the binder; whereas, in the case of embedding, the catalyst would be in direct contact with AP, and thus more effective.

The propellant burn rates could also be increased by having a higher specific surface area of aluminum in the propellant. Nano-sized aluminum has been employed in place of micrometer-sized aluminum and has been reported by Dokhan et al. (2002) to result in very high burn rates in composite solid propellants. But, as the specific surface area of nano aluminum is very large, it increases the viscosity of the propellant slurry drastically and makes it difficult to mix, as observed by Sippel et al. (2013, 2014) and Verma and Ramakrishna (2013). Furthermore, Verma and Ramakrishna (2013) have proposed the use of micrometer-sized flake aluminum (pyral) with a very high surface area and a heat of combustion higher than nano-sized aluminum to increase the burn rates of composite solid propellants.

Apart from the aluminum particle size, fluorination of aluminum has been known to result in very high (volumetric and gravimetric) heat release as compared to its oxidation (Sippel et al. 2013) due to the thermite reactions. Therefore, when aluminum is activated with a fluorine based compound such as polytetrafluoroethylene (PTFE), it results in higher reactivity, thereby increases the burn rates, as reported by Sippel et al. (2014). Similar results were reported by Gaurav and Ramakrishna (2016), where mechanical activation of aluminum with PTFE resulted in a 50% higher burn rate over those obtained by Verma and Ramakrishna (2013) using pyral. In this study, it is intended to make use of mechanically activated aluminum with PTFE for making fuel-rich propellants.

10.5   COMBUSTION STUDIES OF ALUMINIZED FUEL-RICH PROPELLANT

Developing a fuel-rich propellant would be a very challenging task as it involves making a propellant which has no residue upon combustion, high density, high I_{sp} and high burn rates. As the fuel-rich propellant is supposed to burn on its own, it has to have a certain amount of oxidizer present in it for sustained combustion. To achieve zero residue, oxidizer content can be increased, but this will reduce the I_{sp} as inferred from Figure 10.4. The I_{sp} values were obtained with CEA software (Gordon and McBride 1971) using Eq. (10.1) at an altitude of 10 km for a flight Mach number of 3. Figure 10.4 also shows the I_{sp} for the aluminum based fuel-rich propellant developed by Zhongqin et al. (1986). It clearly shows the drop in I_{sp} with their

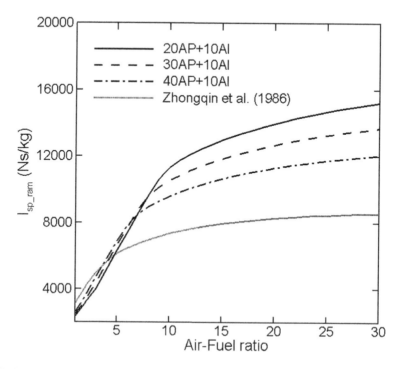

FIGURE 10.4 I_{sp} variation with change in AP percentage.

propellant as it contained 50% by weight of AP. But, if the oxidizer content were to be less, it would result in lot of residue, which again reduces the performance. Similarly, burn rates have to be higher for an end burning configuration (3–5 mm/s). If these burn rates are not achieved the required mass flow rate of fuel can only be attained through a port burning configuration, which will reduce the volumetric loading of the fuel-rich propellant. These are some of the conflicting requirements which makes the development of FRP very difficult. An optimum oxidizer content would be essential to meet all these requirements.

10.6 PROPELLANTS PREPARED

Propellant compositions as shown in Table 10.1 were prepared. These were similar to those prepared by Nanda and Ramakrishna (2013) in terms of binder, aluminum and AP content. However, they differed in the additives added, which were chosen to reduce the residue observed by Nanda and Ramakrishna (2013) and the method used for their inclusion into the propellant has been discussed earlier.

10.7 RESULTS AND DISCUSSION

The Mix 0 shown in Table 10.1 was prepared by Nanda and Ramakrishna (2013), which had a residue of 20–25% and the iron oxide catalyst was added to the propellant and not embedded. Among the propellant Mixes 1 to 5, as shown in Table 10.1, only Mix 4 and 5 had residue of 10–13%; the remaining mixes resulted in a higher residue of the order of 18%. Though the residue of 18% was marginally less than those reported by Nanda and Ramakrishna (2013) of around 20–25%, it was considered high and hence these mixes were not considered for burn rate tests.

It can be seen from Table 10.1 that addition of mechanically activated aluminum in Mix 5 reduced the residue to 10–13% from 18% in Mix 2, where only aluminum was used. This reaffirms the fact that mechanical activation of aluminum with PTFE increases its reactivity, as reported by Sippel et al. (2013, 2014) and Gaurav and Ramakrishna (2016). As stated by Gaurav and Ramakrishna (2016), the thermite

---

TABLE 10.1
Chemical Composition of Fuel-Rich Propellants Prepared.

Mix	AP (wt.%)	IO	AC	Moisture	PTFE	Al	Binder	Residue	Density (kg/m³)
0	30	3	-	-	-	10	57	20–25	1255
1	29.7	-	0.3	0.12	-	10	59.88	18	1200
2	29.7	0.3	-	-	-	10	60	18	1224
3	30	-	-	-	1.765	10	58.235	18	1190
4	29.7	-	0.3	0.12	1.765	10	58.115	10–13	1162
5	29.7	0.3	-	-	1.765	10	58.235	10–13	1212
6	35	-	-	-	-	30	35	0	1528

reaction of aluminum and PTFE results in fluorination of aluminum which releases more energy than oxidation (Sippel et al. 2013). This could be taking place close to the propellant burning surface, and hence result in a large heat feedback to the propellant leading to lower residue. Also, the product of the aluminum PTFE reaction is aluminum fluoride, which sublimates at around 1500 K (Sippel et al. 2014), unlike aluminum oxide, which tends to agglomerate and has a higher boiling point of 3800 K (Athawale et al. 1994).

Also, when AP embedded with AC and moisture absorbed in AC pores was used along with mechanically activated aluminum in Mix 4, it reduced the residue to 10–13% from 18%, where only AP with mechanically activated aluminum is used in Mix 3. This is similar to what was observed between Mix 3 and 5, where IO embedded AP along with mechanically activated aluminum results in lower residue of 10–13%.

A very interesting aspect that comes out of this study is that, when mechanically activated aluminum and catalyst embedded AP (either AC or IO) were used together, the residue was reduced to 10–13%. However, if either one is used independently, the residue was higher (18%). This was something that Nanda and Ramakrishna (2013) had not explored in their studies.

Burn rates measured using Window bomb for Mix 4 and 5 are shown in Figure 10.5. Mix 4 exhibits a higher burn rate pressure index of 0.82, which was quite high to be used in a rocket motor due to fears of stability associated with combustion. The

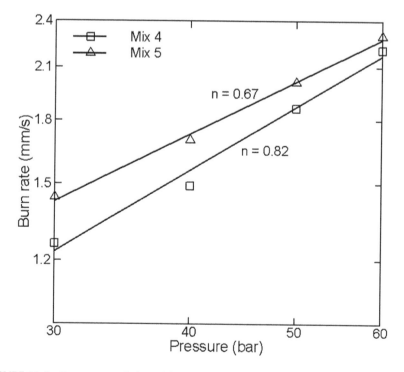

FIGURE 10.5 Burn rate variation with pressure for Mix 4 and 5.

reason for this higher pressure index with Mix 4 had been discussed earlier in conjunction with the work done by Ishitha and Ramakrishna (2014a), where they had reported a higher pressure index due to moisture presence in AC pores. The burn rate pressure index of 0.67 observed in the case of Mix 5 was reasonably good for use in solid fuel ramjets (SFRJ) and provides for a better control for changing the fuel flow rates on command. The highest burn rate recorded was 2.3 mm/s for Mix 5 at 60 bar. The uncertainty in the measurement of the burn rate was around 6.5% with a dispersion within 5%.

The activation of aluminum and AP resulted in the reduction of residue to 10–13% for Mix 4 and 5 compared to Mix 0 developed by Nanda and Ramakrishna (2013). However, the effect of activation of Al and AP on propellant burning is not clear. This calls for certain experiments which could throw more light on the combustion phenomenon involved in burning of these propellants.

The equilibrium temperature obtained using CEA software (Gordon and McBride 1971) in the ramjet primary chamber for these two propellants (Mix 0 and Mix 5) were 1665 K and 1720 K. Although this is a small difference, this increase in equilibrium temperature observed with Mix 5 could be due to the fluorination reaction of aluminum as explained earlier. It is not clear whether this alone could lead to the lower residue noticed with Mix 5.

To understand this thermal behaviour of the fuel-rich propellants, differential scanning calorimetry (DSC) and thermogravimetric analysis (TGA) of Mixes 0 and 5 was carried out in an argon environment for a temperature range of 300 to 1000 K with a heating rate of 40 K per minute in a NETZSCH STA 449F3 instrument. Argon was chosen so as to have an inert environment, which is the case when the fuel-rich propellant burns in the primary chamber. Figure 10.6 and 10.7 shows the results of DSC-TGA analysis for Mix 0 and 5. The onset of the first exothermic peak

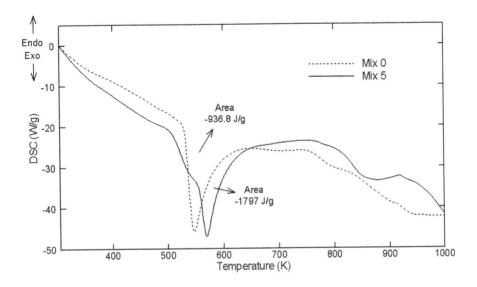

FIGURE 10.6   DSC analysis of Mix 0 and 5.

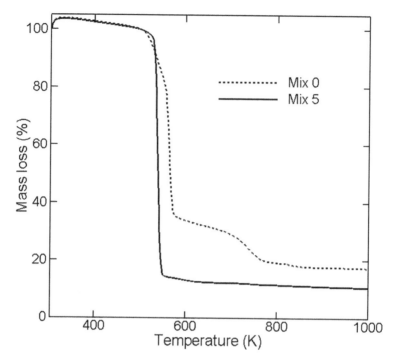

FIGURE 10.7   TGA analysis of Mix 0 and 5.

for Mix 5 was at around 500 K whereas for Mix 0 the onset was at around 520 K as seen in Figure 10.6. The area covered by the first peak was also more for Mix 5, almost 1.9 times that of Mix 0 (refer to Figure 10.6), representing higher heat release and increased reactivity. Figure 10.7 shows the results of TGA analysis, where the loss in mass starts at almost the same temperature for both the propellants; however, the final mass left un-reacted is higher with Mix 0. It can be seen that the drop in mass is steeper in the case of Mix 5 due to its higher reactivity (fluorination reaction of aluminum), which results in faster reaction and conversion of aluminum in solid fuel-rich propellant into hot gases. The final mass left or the residue corresponding to both the mixes is in line with Table 10.1.

The activation of Al and AP has increased the reactivity of the propellant as evident from the DSC analysis, however, what is happening at the propellant burning surface is still unknown. But to observe the propellant burning surface while it is burning is practically not possible. However, if one can stop the propellant combustion rapidly while it is burning, it would give a snapshot of the burning surface at that moment. This can be achieved by quenching the burning propellant by rapid depressurization in a quench bomb setup. The quenched samples of Mix 0 and 5 were analyzed under the scanning electron microscope (SEM) FEI Quanta 200 and the images are as seen in Figure 10.8. The scale shown in the image is for 100 μm. Figure 10.8a shows SEM images of Mix 0 propellant, where binder melt is clearly visible. The propellant surface is covered with large quantities of binder melt.

(a)   (b)

FIGURE 10.8   Images under a SEM at 800 times magnification for (a) Mix 0 and (b) Mix 5.

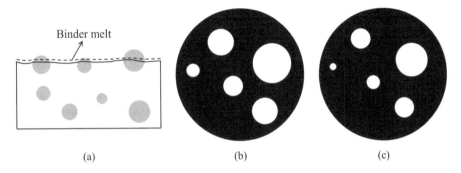

(a)   (b)   (c)

FIGURE 10.9   Representative sketch of a propellant (a) from the side view showing binder melt, (b) top view with no binder melt and (c) top view with binder melt.

Figure 10.9 shows the schematic of a propellant from the side and top view with particles exposed off the propellant surface. As the propellant burns the surface gets covered with binder melt as shown by the dotted line in Figure 10.9a. This reduces the exposure of particles to the flame and completely covers some particles on the surface as seen in Figure 10.9b and 10.9c, thereby restricting the burning of AP and Al resulting in residue and low burn rates. In case of Mix 5, the binder melt is reduced as inferred from Figure 10.8b. This higher exposure of particles allows it to burn better and thereby reduces the residue.

The TGA-DSC, burn rate, residue and SEM images suggest that Mix 5 has lower binder melt than Mix 0. This has been due to the increased reactivity of AP and Al. The PTFE results in aluminum combustion close to the propellant surface as suggested by Sippel et al. (2014) and Gaurav and Ramakrishna (2016), thereby increasing the heat feedback to the propellant. This increases the pyrolysis of the binder resulting in reduced binder melt as seen in Figure 10.8b. Lower binder melt also means more of AP is exposed on the propellant surface as evident from the sketches shown in Figure 10.9. It has been demonstrated by Ishitha and Ramakrishna (2014a)

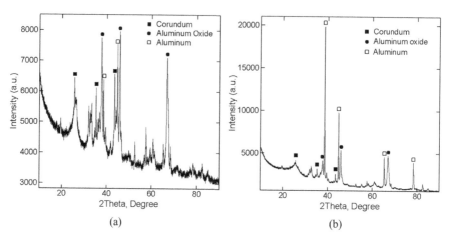

FIGURE 10.10   XRD pattern obtained for (a) Mix 0 and (b) Mix 5.

through careful experiments that as the surface area of AP exposed increases in a solid propellant, correspondingly the pressure index of the propellant also increases. Gaurav and Ramakrishna (2016) have also reported an increase in the pressure index of aluminized solid propellants with PTFE activation of aluminum, which they attribute to the premixedness of aluminum and fluorine. The results presented here are in agreement with earlier literature resulting in a higher burn rate pressure index of 0.67 observed with Mix 5 compared to 0.32 with Mix 0.

The residue obtained after the burning of Mix 0 and 5 was analyzed for its XRD pattern as shown in Figure 10.10. The graph shows the intensity levels at each 2θ position based on the compounds present. The peaks correspond to the compound present at that angle. The XRD data was analyzed using Xpert Highscore for the compounds present in the residue. The intensity level data at each angle is known for each of the compounds. Based on the intensity actually measured at each 2θ position, the presence of that particular compound is shown in Figure 10.10. The points identified are for those cases only where the actual intensity was at least 50% of the intensity data for the corresponding compound. Based on this analysis the weight fractions for each compound were obtained and summarized in Table 10.2. As seen in Table 10.2, the unburnt aluminum content was more for Mix 5; however since the absolute value of residue was less, the amount of unburnt aluminum was same for both the compositions. The oxides of aluminum are much lower for Mix 5 in absolute terms (8.3%) compared to Mix 0 (21.75%). This suggests that in case of Mix 5, Al has both oxygen and fluorine as oxidizers to react and that aluminum has a higher tendency of reacting with the fluorine present in PTFE as it is a better oxidizer than oxygen. The product of this reaction is aluminum fluoride (AlF_3), which sublimates at around 1500 K (Sippel et al. 2014), unlike aluminum oxide, which tends to agglomerate and have a higher boiling point of 3800 K (Athawale et al. 1994). As aluminum combustion in Mix 5 is better it has less residue compared to Mix 0.

With the activation of AP and Al, Mix 5 had a residue of 10–13%, which was still high, making it unsuitable for any practical application. Various other compositions were tried in order to reach a composition with very low or zero residue. It was found that with such high content of binder any further efforts to reduce the residue would be futile as the binder melt would be very high. Therefore, further propellants were made with the solid content in them increased, and a number of iterations led to the development of Mix 6 composition (refer to Table 10.1). An increase in the aluminum content to 30% by weight not only resulted in an increase in propellant density but eliminated the residue completely. The burn rates obtained in Figure 10.11 were higher than those achieved with lower solid loading propellants (Mixes 4 and 5). The uncertainty in the measurement of the burn rate was around 6.5% with a dispersion within 5%.

TABLE 10.2
Weight Fractions of Major Compounds in the Residue.

	Aluminum	Corundum (Al_2O_3)	Aluminum Oxide ($Al_{10.66}O_{16}$)
Mix		(wt. %)	
Mix 0	6	46	41
Absolute value for Mix 0	1.5	11.5	10.25
Mix 5	14	14	69
Absolute value for Mix 5	1.4	1.4	6.9

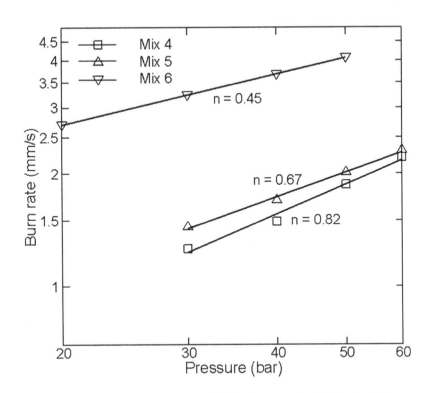

FIGURE 10.11   Burn rate versus pressure for Mix 6 compared with Mix 4 and 5.

FIGURE 10.12   SEM image of a quenched sample of Mix 6 with no visible binder melt.

In order to understand how increasing the aluminum content led to a zero residue fuel-rich propellant, the quenched sample of Mix 6 was observed under SEM. From Figure 10.12 it is evident that the quenched surface shows very little signs of binder melt. Due to higher aluminum content in Mix 6, the equilibrium temperature obtained using CEA software (Gordon and McBride 1971) in the ramjet primary chamber was around 2410 K, much higher compared to those obtained with Mix 0 and 5. The resultant heat feedback could have resulted in more complete pyrolysis of the binder, resulting in nearly no visible binder melt on the propellant burning surface.

10.8   CONCLUSION

Methods based on literature for increasing the reactivity of AP and aluminum were applied to the propellants that were prepared with higher binder content. The residue was reduced from 20–25% to 10–13%. The highest burn rate for these propellants was around 2.3 mm/s for Mix 5 at 60 bar. The activation resulted in burning occurring close to the propellant surface which ensured higher pyrolysis of binder (as found in TGA-DSC and SEM analysis). However, activation alone could not bring the residue to zero. The SEM images of the quenched samples showed a reduction in binder melt which helped in reducing the residue to 10–13% as more particles were exposed at the propellant burning surface. This reduction in melt was attributed to the increased reactivity of AP and aluminum. Further propellants made with the similar solid loading did not result in any further reduction in residue. It was found that an aluminum content of 30% by weight would result in a propellant with zero residue (Mix 6). It had the highest burn rate of around 4 mm/s at 50 bar. Along with zero residue it had an added advantage of higher density of 1528 kg/m^3. The SEM images of the quenched sample showed nearly no binder melt on the propellant surface which helped in eliminating the residue.

REFERENCES

Athawale, B., S. Asthana, and H. Singh (1994). Metallised fuel-rich propellants for solid rocket ramjet-a review. *Defence Science Journal*, 44(4), 269–278.

Dokhan, A., E. W. Price, J. M. Seitzman, and R. K. Sigman (2002). The effects of bimodal aluminum with ultrafine aluminum on the burning rates of solid propellants. *Proceedings of the Combustion Institute*, 29(2), 2939–2946.

Gany, A. (2006). Effect of fuel properties on the specific thrust of a ramjet engine. *Defence Science Journal*, 56(3), 321–328.

Gany, A. and D. Netzer (1985). Fuel performance evaluation for the solid-fueled ramjets. *International Journal of Turbo and Jet Engines*, 2, 157–168.

Gany, A. and D. Netzer (1986). Combustion studies of metallized fuels for solid-fuel ramjets. *Journal of Propulsion and Power*, 2(5), 423–427.

Gaurav, M. and P. A. Ramakrishna (2016). Effect of mechanical activation of high specific surface area aluminium with PTFE on composite solid propellant. *Combustion and Flame*, 166(2), 203–215.

Gobbo-Ferreira, J., M. Silva, and J. Carvalho Jr (1999). Performance of an experimental polyethylene solid fuel ramjet. *Acta Astronautica*, 45(1), 11–18.

Gordon, S. and B. McBride (1971). Computer program for calculation of complex chemical equilibrium compositions, rocket performance, incident and reflected shocks, and Chapman-Jouguet detonations. Technical Report NASA SP-273, NASA Lewis Research Center; Cleveland, OH, United States.

Ishitha, K. and P. A. Ramakrishna (2014a). Activated charcoal: As burn rate modifier and its mechanism of action in nonmetallized composite solid propellants. *International Journal of Advances in Engineering Sciences and Applied Mathematics*, 6(1), 76–96.

Ishitha, K. and P. A. Ramakrishna (2014b). Studies on the role of iron oxide and copper chromite in solid propellant combustion. *Combustion and Flame*, 161, 2717–2728.

Kohga, M. (2011). Burning characteristics and thermochemical behavior of AP/HTPB composite propellant using coarse and fine ap particles. *Propellants, Explosives, Pyrotechnics*, 36(1), 57–64.

Krishnan, S. and P. George (1998). Solid fuel ramjet combustor design. *Progress in Aerospace Sciences*, 34, 219–256.

Kubota, N. and T. Kuwahara (1991). Combustion of energetic fuel for ducted rockets (i). *Propellants, Explosives, Pyrotechnics*, 16(2), 51–54.

Kubota, N., K. Miyata, T. Kuwahara, M. Mitsuno, and I. Nakagawa (1992). Energetic solid fuels for ducted rockets (iii). *Propellants, Explosives, Pyrotechnics*, 17(6), 303–306.

Kubota, N., N. Y. Yano, K. Miyata, T. Kuwahara, M. Mitsuno, and I. Nakagawa (1991). Energetic solid fuels for ducted rockets (ii). *Propellants, Explosives, Pyrotechnics*, 16(6), 287–292.

Marothiya, G., K. Ishitha, and P. A. Ramakrishna (2013). A new and effective method to enhance the burn rate of composite solid propellants. In 9th Asia-Pacific Conference on Combustion. Gyeongju, Korea.

Nanda, J. K. and P. A. Ramakrishna (2013). Development of AP/HTPB based fuel-rich propellant for solid propellant ramjet. In 49th AIAA/ASME/SAE/ASEE Joint Propulsion Conference and Exhibit (JPC). San Jose, CA.

Natan, B. and A. Gany (1993). Combustion characteristics of a boron-fueled solid fuel ramjet with aft-burner. *Journal of Propulsion and Power*, 9(5), 694–701.

Raghunandan, B. N., E. Ravichandran, and A. Marathe (1985). Combustion related to solid-fuel ramjets. *Journal of Propulsion and Power*, 1(6), 502–504.

Risha, G. A., B. J. Evans, E. Boyer, and K. K. Kuo (2007). Metals, energetic additives and special binders used in solid fuels for hybrid rockets. In *Fundamentals of Hybrid Rocket Combustion and Propulsion*, volume 218. Progress in Astronautics and Aeronautics, AIAA, Reston, VA.

Schulte, G. (1986). Fuel regression and flame stabilization studies of solid-fuel ramjets. *Journal of Propulsion and Power*, 2(4), 301–304.

Sippel, T. R., S. F. Son, and L. J. Groven (2013). Altering reactivity of aluminum with selective inclusion of polytetrafluoroethylene through mechanical activation. *Propellants, Explosives, Pyrotechnics*, 38(2), 286–295.

Sippel, T. R., S. F. Son, and L. J. Groven (2014). Aluminum agglomeration reduction in a composite propellant using tailored Al/PTFE particles. *Combustion and Flame*, 161(1), 311–321.

Verma, S. and P. A. Ramakrishna (2013). Effect of specific surface area of aluminium on composite solid propellant burning. *Journal of Propulsion and Power*, 29(5), 1200–1206.

Zhongqin, Z., Z. Zhenpeng, T. Jinfu, and F. Wenlan (1986). Experimental investigation of combustion efficiency of air-augmented rockets. *Journal of Propulsion and Power*, 2(4), 305–310.

Index

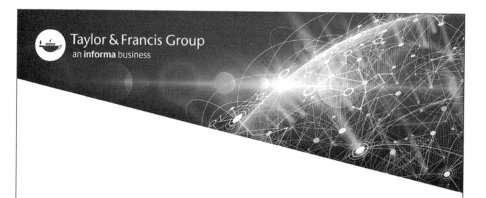

For Product Safety Concerns and Information please contact our
EU representative GPSR@taylorandfrancis.com Taylor & Francis
Verlag GmbH, Kaufingerstraße 24, 80331 München, Germany